SpringerBriefs in Molecular Science

Chemistry of Foods

Series editor

Salvatore Parisi, Industrial Consultant, Palermo, Italy

The series Springer Briefs in Molecular Science: Chemistry of Foods presents compact topical volumes in the area of food chemistry. The series has a clear focus on the chemistry and chemical aspects of foods, topics such as the physics or biology of foods are not part of its scope. The Briefs volumes in the series aim at presenting chemical background information or an introduction and clear-cut overview on the chemistry related to specific topics in this area. Typical topics thus include: - Compound classes in foods – their chemistry and properties with respect to the foods (e.g. sugars, proteins, fats, minerals, …) - Contaminants and additives in foods – their chemistry and chemical transformations - Chemical analysis and monitoring of foods - Chemical transformations in foods, evolution and alterations of chemicals in foods, interactions between food and its packaging materials, chemical aspects of the food production processes - Chemistry and the food industry – from safety protocols to modern food production The treated subjects will particularly appeal to professionals and researchers concerned with food chemistry. Many volume topics address professionals and current problems in the food industry, but will also be interesting for readers generally concerned with the chemistry of foods. With the unique format and character of Springer Briefs (50 to 125 pages), the volumes are compact and easily digestible. Briefs allow authors to present their ideas and readers to absorb them with minimal time investment. Briefs will be published as part of Springer's eBook collection, with millions of users worldwide. In addition, Briefs will be available for individual print and electronic purchase. Briefs are characterized by fast, global electronic dissemination, standard publishing contracts, easy-to-use manuscript preparation and formatting guidelines, and expedited production schedules. Both solicited and unsolicited manuscripts focusing on food chemistry are considered for publication in this series.

More information about this series at http://www.springer.com/series/11853

Angela Montanari · Caterina Barone
Michele Barone · Anna Santangelo

Thermal Treatments of Canned Foods

 Springer

Angela Montanari
Stazione Sperimentale per L'Industria delle
  Conserve Alimentari, Food Packaging
Parma
Italy

Michele Barone
Associazione 'Componiamo il Futuro'
  (CO.I.F.)
Palermo
Italy

Caterina Barone
Associazione 'Componiamo il Futuro'
  (CO.I.F.)
Palermo
Italy

Anna Santangelo
Food Technologist
San Marco Evangelista, Caserta
Italy

ISSN 2191-5407 ISSN 2191-5415 (electronic)
SpringerBriefs in Molecular Science
ISSN 2199-689X ISSN 2199-7209 (electronic)
Chemistry of Foods
ISBN 978-3-319-74131-4 ISBN 978-3-319-74132-1 (eBook)
https://doi.org/10.1007/978-3-319-74132-1

Library of Congress Control Number: 2017963529

Printed on acid-free paper

This Springer imprint is published by the registered company Springer International Publishing AG part of Springer Nature
The registered company address is: Gewerbestrasse 11, 6330 Cham, Switzerland

Contents

Chapter 1
Canned Foods: Principles of Thermal Processing

Angela Montanari, Caterina Barone, Michele Barone
and Anna Santangelo

Abstract The history of food industry is strictly correlated with a peculiar category of long-durability edible products: canned foods. Differently from other packaged foods, canned foods show several unique properties, including risks and failures, depending on the composition of edible contents, the production of metal packages and preservation techniques. Thermal processes have the basic aim of destroying microorganisms (bacteria and spore-forming life forms) in foods. The inhibition of microbial growth and the inactivation of microbial toxins are also needed. Other factors—pH, presence of fatty molecules, calcium, etc—are important. As a result, the choice of the 'right' thermal treatment (pasteurisation, sterilisation) and related process parameters (time, temperature) have to be considered in different ambits, including canned foods (Chen in J Food: Microbiol Saf Hyg 02(1), 2017; Chen et al. in J Sci Food Agric 93(5):981–986, 2013; Gupta and Balasubramaniam in Novel thermal and non-thermal technologies for fluid foods. Academic Press, London, Waltham, and San Diego, pp. 109–133, 2012; IciEr in Novel thermal and non-thermal technologies for fluid foods. Academic Press, London, Waltham, and San Diego, pp. 305–367, 2012; Jongyingcharoen and Ahmad in Functional foods and dietary supplements: processing effects and health benefits. Wiley, Chichester, UK, 2014; Mañas and Pagán in J Appl Microbiol 98(6):1387–1399, 2005; Rastogi in Novel thermal and non-thermal technologies for fluid foods. Academic Press, London, Waltham, and San Diego, pp. 411–432, 2012; Sahin and Sumnu in Physical properties of foods, pp. 107–155, 2006; Tiwari and Mason in Novel Thermal and non-thermal technologies for fluid foods. Academic Press, London, Waltham, and San Diego, pp. 135–165, 2012; Vasseur et al. in J Appl Microbiol 86 (3):469–476, 1999). Final results are the construction of logarithmic 'survival curves', the definition of factors which can reduce thermal destruction of microorganisms (water activity and pH). In addition, some reflection should be made when speaking of 'commercial sterility' and the correlated concept of long durability for canned foods (storage at room temperature).

Keywords Acidity · Canned food · Commercial sterility · Decimal reduction time
Mycotoxin · Survival curve · Thermal treatment

© The Author(s), under exclusive licence to Springer International Publishing AG, part of Springer Nature 2018
A. Montanari et al., *Thermal Treatments of Canned Foods*,
Chemistry of Foods, https://doi.org/10.1007/978-3-319-74132-1_1

Abbreviations

D_T Decimal reduction time

1.1   Thermal Food Processing: The History of Food Industry

The history of food industry is strictly correlated with a peculiar category of long-durability edible products: canned foods. Differently from other packaged foods, canned foods show several unique properties, including risks and failures, depending on the composition of edible contents, the production of metal packages and preservation techniques. In particular, thermally preserved canned foods need to be studied with a certain attention because of the critical importance of thermal treatments in terms of processing parameters and desired results (Chen 2017; Chen et al. 2013; Gupta and Balasubramaniam 2012; Jongyingcharoen and Ahmad 2014; Mañas and Pagán 2005; Rastogi 2012; Sahin and Sumnu 2006; Tiwari and Mason 2012; Vasseur et al. 1999).

Generally, canned foods can undergo treatments such as pasteurisation and sterilisation depending of the contained product, the related microbial ecology, extended durability requirements (shelf-stable food products) and certain packaging-related features such as heat transfer (Mauer and Ozen 2004). Thermal processes have the basic aim of destroying microscopic life forms (bacteria and spore-forming organisms) in foods; the inhibition of microbial growth and the inactivation of microbial toxins are also needed. Other factors—pH, presence of fatty molecules, calcium, etc.—have to be considered because thermal resistance of certain microorganisms can be increased in these conditions. As a result, the choice of the 'right' thermal treatment (pasteurisation, sterilisation) and related process parameters (time, temperature) have to be considered on the basis of a multifaceted analysis, including kinetic considerations (Brown 2000; Carslaw and Jaeger 1959; Engehnan and Sani 1983; Hersom and Hulland 1981; Jen et al. 1971; Lenz and Lund 1977; Lund 1982; Mishkin et al. 1982; Pflung and Odlaugh 1978; Teixeira 1971).

Thermal processing is the technological answer to the resistance of microbiological agents that are able to affect human health and cause significant food degradation. As a result, the current situation and knowledge of thermal processing in the industry have to be considered as the heritage of the first works on this argument, carried out by Estey and Meyer in 1922 (Larousse and Brown 1997). Actually, the basis of thermal processing should be named 'thermobacteriology'. In fact, thermal destruction—the technological strategy that is able to reduce and eliminate microbial activity by heat—cannot be reliably applied without a strong knowledge of thermal resistance ascribed to microorganisms (Ball and Olson 1957; Doyle and Mazzotta 2000; Pflug et al. 2001; Stumbo 1973). The final results of thermal resistance studies are logarithmic 'survival curves' (Peleg 2003; Peleg and

Cole 1998; Peleg et al. 2005): these mathematical representations of microbial evolutive profiles depend on various factors that are able to reduce—or enhance—thermal destruction of microorganisms (water activity, pH, etc.). In addition, some reflection may be explained when speaking of 'commercial sterility' (Schmitt 1966) and the correlated concept of long durability for canned foods (storage at room temperature).

The discussion of evolutive profile of selected microbial agents is not the aim of this chapter; on the other hand, thermal processing techniques depend historically on past studies such as early Appert's procedures for preserving perishable edible products in glass containers by heat penetration (Stumbo 1973). Subsequently, different researchers such as Pasteur, Underwood and Prescott discovered new laws of thermal processing. The introduction of the so-called 'sanitary' can, the use of the first thermocouple for measure purposes during heat preservation experiments by researchers such as Bigelow and Ball in the 1920s, and the creation of 'High-temperature Short-Time' processes led to the powerful progress in thermal preservation (Stumbo 1973).

Anyway, thermal resistance of microorganisms and thermal food processing are interconnected matters. The key junction points are represented by chemical and physical conditions for the detailed knowledge of microbial ecology in selected foods on the one side, and the optimal thermal destruction of microbial agents on the other hand. The following sections are dedicated to the description of these basic features, while Sect. 1.4 tries to explain the concept of 'commercial sterility' with a reliable interpretation of basic facts and practical knowledge from the 'industrial' viewpoint.

1.2   Thermal Food Processing: Microbiological Agents

1.2.1   The Microbial 'Enemy'

The active 'subject' (or the 'enemy') of thermal processing is always a microbial agent (Sect. 1.3). Actually, this statement should be discussed in detail and 'broadened' because of the existence of different 'agents' that are able to cause damages, and related differences between thermal resistances (Larousse and Brown 1997). As a result, microbiological 'agents' should be considered as follows (Jay et al. 2008a, b, c):

(a) Viruses;
(b) Bacteria, including spore-forming microorganisms and
(c) Fungi.

In addition, the third microbiological category is able to produce toxins with notable heat resistance. Actually, some bacteria can also produce staphylococcal and botulinal toxins, but the related thermal inactivation appears to be relatively easy. On the other side, certain enterotoxins by *Staphylococcus aureus* are reported to remain active after thermal processing provided that their production has been observed before canning (Bennett and Berry 1987; Dangerfield 1973; Larousse and Brown 1997; NFPA/CMI Container Integrity Task Force 1984).

With reference to viruses, the main danger appears related to the nucleic acids contained in the organic structure after protein denaturation. In detail, protein-made capsids containing the viral genetic code are rapidly denatured at 50–70 °C, but viral nucleic acids—the real danger when speaking of viruses—may remain active biologically unless a drastic treatment is performed. However, similar treatments—10 min at 100 °C or more in a moist environment can be sufficient—are practically 'implicit' and assumed when speaking of common thermal processing for low-acid canned foods (Blackwell et al. 1985; Charm and Landau 1987; Gaze 2005; Jay et al. 2008a; Larousse and Brown 1997; Richards et al. 2010).

Bacteria are a different 'enemy' because several types are able to produce spores. It may be assumed that canning industry has 'born' as the answer to the survival of dangerous microbial species that are able to produce spores after the inhibition of vegetative life and spreading, after a 100 °C-treatment for several hours (Carlin 2011; Couvert et al. 2005; Larousse and Brown 1997; Oomes et al. 2007). The problem depends on the composition of endospores, a sort of dehydrated structure containing only basic cellular constituents for life (Atrih and Foster 1999; Gates et al. 2010; Henriques and Moran 2000; Jay et al. 2008b, c; McKenney et al. 2013). This strategy allows certain bacteria to assure their 'dormant' presence in adverse conditions without vegetative growth; subsequently, endospores may develop their 'active' or vital form when conditions assure a more favourable environment, with the needed bioavailability of basic elements such as carbon, phosphorus and nitrogen (Atrih and Foster 1999; Moir 2006; Paidhungat and Setlow 2002; Wells-Bennik et al. 2016).

Several spore-forming species are considered when speaking of thermal processing in canning industries: *Bacillus, Clostridium, Actynomycetes, Thermoactinomyces, Pasteuria, Metabacterium,* etc. (André et al. 2013; Dotzauer et al. 2002; Jay et al. 2008b; Matsuda et al. 1985; Pflug et al. 2001; Richardson 1972). Actually, a critical and basic factor affecting the production of spores is the peculiar microbial behaviour: obligate aerobes such as *Bacillus subtilis* can be able to produce spores in a controlled environment (nutrient agar media) within four days at 27–30 °C (Jay et al. 2008d). On the other side, facultative anaerobe species such as *B. coagulans* need at least 50 °C in a dedicated medium for a similar spore-forming performance. On the other side, *Clostridium botulinum* is reported to produce spores at 28–30 °C but related performances in a controlled medium are notably slower than *B. subtilis*: 7–14 days (Larousse and Brown 1997).

Naturally, these performances can be explained in terms of temperature, pH (5.4–5.7 is the optimal range for *B. pasterianum*, while *B. stearothermophilus* should need neutral pH values), aerobic or anaerobic environments, bioavailability of certain ions (calcium, zinc, sulphate and phosphate ions), sugars and nitrogen-based molecules, etc. The synergic action of these factors could explain the specific behaviour for these bacteria when speaking of food products and other non-food ambits (Ahmed and Dirar 2011; Kihm et al. 1988; Labbe and Duncan 1974; Larousse and Brown 1997; Mazas et al. 1997). Anyway, sporulation is always dependent on one or more stress factors, including notable pH variations, thermal shock or the presence of peculiar sulphur-containing molecules (Casolari 1996a; Larousse and Brown 1997).

Thermal resistance of bacterial spores is naturally correlated with physical and chemical features (Jay et al. 2008e); as a result, low pH values can notably increase spore damages under drastic heat conditions. This strategy is particularly useful when speaking of canned foods with low acidity values, although obtained performances are not always guaranteed. On the other side, thermal resistance may increase depending on several factors including the availability of long-chain fatty acids, peculiar metals (iron and calcium), relatively low water activity and the prolonged (non-lethal) exposure to temperatures between 63 and 100 °C (Larousse and Brown 1997).

It should also be noted that fungal spores do not show notable problems if compared with bacterial spores: in particular, the presence of fungi may represent safety and quality risks when speaking of non-sterilised products only. In other words, the main problem of moulds and yeasts is generally correlated with the production of toxins, and this problem is discussed when speaking of spreading bacteria (Larousse and Brown 1997).

The danger represented by bacterial and fungal toxins is different depending on their typology. Botulinal and staphylococcal toxins can be a problem if their inactivation is not observed in normal heat treatments for canned foods, and these conditions seem sufficient for this purpose, although some critical risk may be reported (Dangerfield 1973; Jay et al. 2008d, f; Larousse and Brown 1997; NFPA/ CMI Container Integrity Task Force 1984). On the other side, mycotoxins can represent a serious problem because of their diversification (producing species: *Aspergillus, Penicillium, Cladosporium, Alternaria, Fusarium*, etc.), their different chemical structure, the relationship between toxinogenesis for a selected toxin and the peculiar strain and enhanced thermal resistances. Three examples can be useful (Kitabatake et al. 1991; Larousse and Brown 1997):

(a) Aflatoxin B_1 (Fig. 1.1) can be inactivated at 267 °C minimum (atmospheric pressure);
(b) Citrinin (Fig. 1.2) is thermoinactivated at 160–175 °C depending on conditions and
(c) Fumonisin B_1 (Fig. 1.3) may be inactivated at 150–200 °C.

The 'mycotoxin risk' is dependent on various factors (Jay et al. 2008f), but the main factor of these variables is always the improper storage of raw materials, especially vegetables and fruits, for long periods. Consequently, thermal treatments might have a meaning of 'technical remediation' of canned products, but their efficacy is not absolutely guaranteed because of the notable thermal resistance of toxins. Actually, these molecules should be considered part of the 'chemical risk'

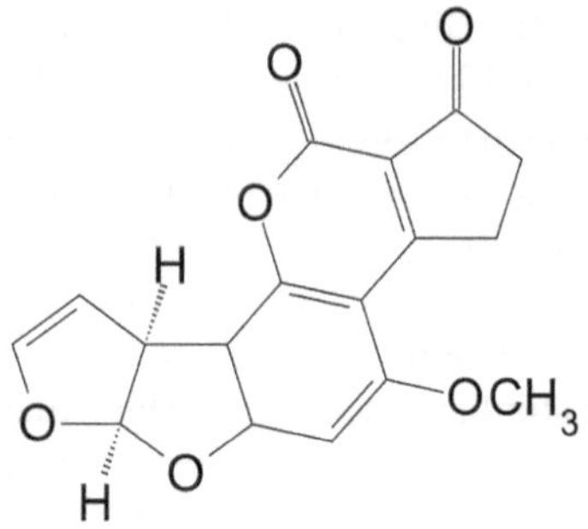

Fig. 1.1 The chemical structure of aflatoxin B$_1$. This mycotoxin) can be inactivated at 267 °C minimum (atmospheric pressure). BKchem version 0.13.0, 2009 (http://bkchem.zirael.org/index. html) has been used for drawing this structure

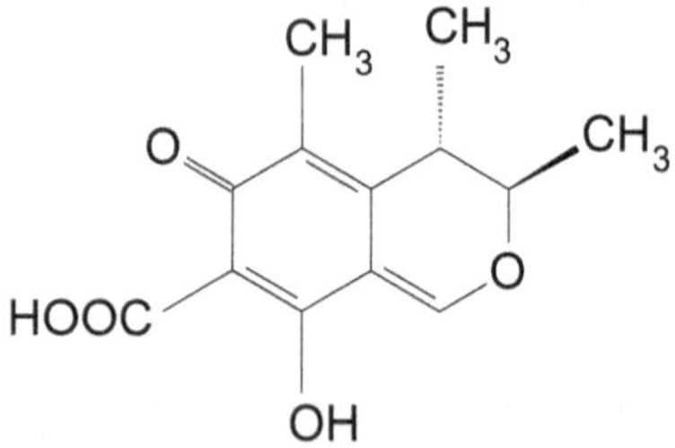

Fig. 1.2 The chemical structure of citrinin. This mycotoxin can be thermoinactivated at 160–175 °C depending on conditions. BKchem version 0.13.0, 2009 (http://bkchem.zirael.org/index. html) has been used for drawing this structure

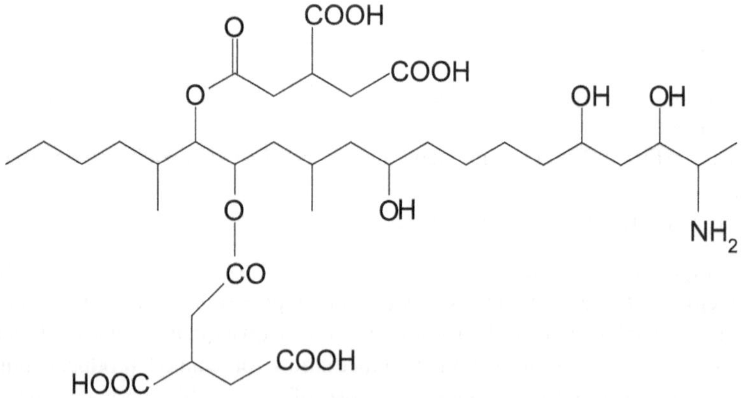

Fig. 1.3 The chemical structure of fumonisin B$_1$. This mycotoxin may be inactivated at 150–200 °C. BKchem version 0.13.0, 2009 (http://bkchem.zirael.org/index.html) has been used for drawing this structure

area; however, they are generally discussed in relation to microbiological risks because of their clear microbiological origin.

1.2.2   Thermal Survival Curves

The science of thermal treatments in the industry has obtained extremely useful results by means of the definition of thermal survival curves (Casolari 1996b; Gómez-Sánchez 2007; Jay et al. 2008e; Larousse and Brown 1997; Schmitt 1966; Setlow and Johnson 2013). The theoretical discussion of survival curves is not considered here; however, it can be affirmed that the comprehension of thermal treatments is strongly linked with the behaviour of living microorganisms under drastic (lethal) conditions. In general, the decrease of living microorganisms under drastic conditions can be expressed by means of the following Eq. 1.1:

$$\log N = \log N_0 - K \times t, \tag{1.1}$$

where N is the number of living microorganisms after a specified thermal treatment for a supposed lethal temperature, N_0 corresponds to the initial number of living microorganisms before treatment, K is a peculiar constant (inactivation speed) for this treatment and t means the treatment time.

This equation is used to determine the decimal reduction time D_T. In detail, should N_2 be the number of living microorganisms survived at the time t_2 for a specified lethal temperature T, and N_1 be the initial number—10 times higher than N_2—at the time t_1, the decimal reduction time would be calculated as follows (Eq. 1.2):

$$D_T = \frac{1}{K} = (t_2 - t_1)/(\log N_1 - \log N_2) \tag{1.2}$$

In other words, D_T represents the thermal resistance for a specified microorganism and a lethal temperature; D_T depends on time only and corresponds to the temporal interval needed to inactivate 90% of initial living microorganisms. Consequently, the higher this number, the higher the thermal resistance for a specified microorganism (Casolari 1996b). Survival curves may be represented by nonlinear equations: this behaviour is observed mainly with relation to vegetative cells, while spores appear to be more 'regular' by this viewpoint (Larousse and Brown 1997).

In addition, D_T is dependent on temperature (for a specified microorganism). By the mathematical viewpoint, Eq. 1.3 shows the relation between D_{T2} and D_{T1} values on the one side and correlated T_1 and T_2:

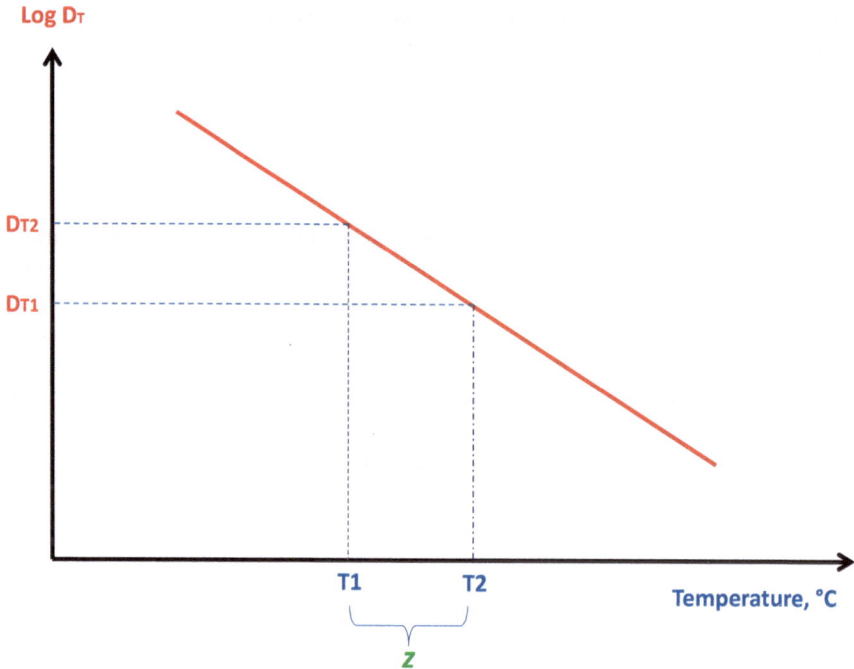

Fig. 1.4 D_T in function of thermal values for a specified microorganism. Equation 1.3 shows the relation between D_{T2} and D_{T1} values on the one side and correlated T_1 and T_2 values. z corresponds to the slope of the linear equation obtained with the representation of D_T values (dependent variable, Y-axis) as logarithms against T values (X-axis)

$$\log D_{T1} - \log D_{T2} = \frac{(T_2 - T_1)}{z},\tag{1.3}$$

where z corresponds to the slope of the linear equation obtained with the representation of D_T values (dependent variable, Y-axis) as logarithms against T values (X-axis). Figure 1.4 displays the situation.

Because of the meaning of D_T values, z means the difference between two specified temperatures, from T_1 to T_2, for the desired 90%-reduction of D_T values, from D_{T1} to D_{T2} (Casolari 1996b). As a result, D_T and z can be obtained and the strategy for thermal treatments is decided for a specified microorganism and the related canned food (Larousse and Brown 1997).

1.2.3 Microbial Alterations: Main Degradation Effects

Different microbial contamination episodes may determine different visible alterations, but normal consumers do not always observe microbial spreading. In general, the following contamination signs are revealed depending on the peculiar microbial 'enemy' in non-acid products with pH > 5.3 (AACC International 2009; Blaschek 1999; Casolari 1996a; Evancho and Walls 2001; Fields 1970; Gooch 2011; Johnson 1999; McClane 2007; Miyamoto and Nagahama 2016; Notermans 1999; Olson and Sorrells 2015):

(1) Flat sour effect. It means the progressive acidification of carbohydrates with lactic acid production and without a notable volumetric increase of containers. pH values are reported to decrease up to 1 log units, with degraded product features (abnormal aroma and taste). This defect is generally ascribed to *B. stearothermophilus* because of insufficient sterilisation and a measurable *Bacillus* contamination. Actually, a minor sign of contamination is reported: the diminution of inner pressure into containers and

(2) Putrefaction. This phenomenon can be ascribed to *C. sporogenes, C. botulinum, C. perfringens* and other Clostridia. In general, pH values do not seem modified, although some increase has been reported. Containers are always deformed (with explosion risks) because of remarkable gas production: hydrogen and carbon dioxide. Sensorial features may be modified when speaking of *C. sporogenes* or other mesophilic Clostridia (production of hydrogen sulphide, organic tiols, etc.), while *C. botulinum* and *C. perfringens* may degrade foods with little organoleptic modifications. C. botulinum contamination can be lethal. Anyway, sterilisation is insufficient. Several episodes are ascribed to anaerobic and thermophilic Clostridia such as *C. thermosaccharolyticum*, with notable modifications of foods and/or containers.

Interestingly, the explosion of containers can be easily observed at 30 °C (exception: flat sour defects). On the other side, acid food products (pH < 4.5) and medium-acidity foods (pH between 4.5 and 5.3) can be degraded in the following ways (Larousse and Brown 1997; Tucker and Featherstone 2011):

(a) Flat sour by *B. coagulans*, with little sensorial modifications. In general, preserving treatments have been insufficient;

(b) pH lowering without container deformations, by Lactobacillaceae. Food recontamination after preserving treatment is the common cause for this defect and

(c) Gas production with container deformation and possible explosion, by *C. pasteurianum* (production of hydrogen, carbon dioxide, butyric acid), yeasts (production of carbon dioxide and ethanol), *Leuconostoc mesenteroides* and *L. dextranicum* (production of carbon dioxide and biological polymers), moulds (production of carbon dioxide; evident mycelia; toxinogenesis) and several Lactobacillaceae (production of lactic acid, carbon dioxide, ethanol, etc.). Different technological causes may be considered, but insufficient thermal treatments and recontamination (because of non-hermetic containers) have to be evaluated.

1.3   Thermal Food Processing: Chemical and Physical Variables Affecting Thermal Resistance

D_T and z values depend on the peculiar temperature, the specified microorganism and several chemical–physical variables (Sect. 1.2). Low pH values can notably increase spore damages under drastic heat conditions, in general, and the same behaviour has been reported when low water activity values influence negatively thermal destruction. On the other side, thermal resistance may increase depending on several factors including the availability of long-chain fatty acids, peculiar metals (iron and calcium) and the prolonged (non-lethal) exposure to temperatures between 63 and 100 °C.

In detail, thermal resistance increases if water activity decreases, and this situation has been observed mainly when speaking of vegetative cells. By the mathematical viewpoint, it can be assumed in general (vegetative cells and spores) that D_T becomes 10 times higher if water activity decreases of 0.2–0.3 units, while an opposite behaviour may be observed if water activity is <0.3 (Casolari 1996b). At the same time, z values tend to increase notably. On the other hand, spores are certainly more resistant than vegetative cells because their structure contains low water amounts.

Interestingly, environmental conditions may play an important role when speaking of supports (apparently, plastic surfaces may increase D_T values in comparison with metal surfaces such as stainless materials and papers), presence of fat matter in the medium (D_T is high in pure oils, while water/oil emulsions seem to reduce thermal resistance similarly to normal aqueous solutions) and air speed (air circulation: the higher the speed, the lower the experimental D_T) (Casolari 1996b). Many other variables should be considered in different ambits when speaking of microbial inactivation and destruction by means of thermal strategies (Chen 2017; Chen et al. 2013; Gupta and Balasubramaniam 2012; Jongyingcharoen and Ahmad 2014; Mañas and Pagán 2005; Rastogi 2012; Sahin and Sumnu 2006; Tiwari and Mason 2012; Vasseur et al. 1999).

With relation to acidity, pH values can increase or decrease thermal resistance depending on involved microorganisms. In general, bacteria are more or less easily destroyed at neutral pH values, while moulds and yeasts appear to resist well at pH < 5.5. However, a general rule does not exist although some approximate relationship between pH and D_T values has been proposed. The same thing can be affirmed when speaking of z values (Casolari 1996b).

The negative influence of certain metals such as iron may be considered. The prolonged and non-lethal exposure to certain temperatures <100 °C is generally observed when speaking of dormant spores produced by *C. botulinum* and *B. stearothermophilus*, with a possible explanation related to the loss of enzymatic activity (Etoa and Michiels 1988).

1.4 The 'Commercial Sterility' Concept

The 'commercial sterility' concept is often proclaimed and repeated (or accepted) in the food industry and in commercial sectors (Augusto et al. 2014; Costa 2003; Da Silva et al. 2013; Hall 1971; Heinz and Hautzinger 2007; Jay et al. 2008e; Membré and van Zuijlen 2011; Nightingale and Stallings 1986; Schmitt 1966). However, the theory of thermal destruction demonstrates that the survival probability of living microorganisms cannot be reduced to the 'zero' value. In other terms, 'absolute sterility' is a concept without mathematical and reliable bases (Casolari 1996b). Moreover, the construction of thermal destruction time curves could not be possible on these bases because of the logarithmic nature of these graphs. In addition, the difference between N and N_0 values corresponds to the amount of killed microorganisms, where 'killed' or 'dead' microorganism is defined in this way because of the impossible reproductive ability (Larousse and Brown 1997; Stumbo 1973). Actually, a certain fraction of 'dead' spores could still be surviving but unable to show normal metabolic functions unless culture conditions become more favourable.

As a result, the term 'commercial sterility' should be correctly defined, in particular when speaking of low-acid canned foods. In this ambit, canned foods are 'commercially sterile' after a heat treatment that is able to destroy pathogen and toxin-producing microorganisms, including also resistant microbiological agents which could be able to spread in the canned food under normal storage conditions (Brooks and Houpt 1975). In these conditions, 'commercially sterile' canned foods are expected to remain stable for extended storage periods, although non-stable products may be found on the market (André et al. 2013). On the other hand, the term 'absolute sterility' has not practical meanings because of its practical impossibility (Von Bockelmann and Von Bockelmann (1986).

References

AACC International (2009) Thermophilic spore counts (Total Aerobic, Flat-Sour, H_2S, Non-H_2S Anaerobic). Approved methods of analysis, 11th Ed. Method 42-40-01. American Association of Cereal Chemists (AACC) International, St. Paul. https://doi.org/10.1094/aaccintmethod-42-40.01

Ahmed AA, Dirar HA (2011) Effect of aeration and method of addition of glucose sugar to culture medium on growth and sporulation of some *Bacillus thuringiensis* isolates from Sudan soils. Gezira J Agric Sci 9:1

André S, Zuber F, Remize F (2013) Thermophilic spore-forming bacteria isolated from spoiled canned food and their heat resistance. Results of a French ten-year survey. Int J Food Microbiol 165(2):134–143. https://doi.org/10.1016/j.ijfoodmicro.2013.04.019

Atrih A, Foster SJ (1999) The role of peptidoglycan structure and structural dynamics during endospore dormancy and germination. Antonie Van Leeuwenhoek 75(4):299–307. https://doi.org/10.1023/A:1001800507443

Augusto PED, Tribst AAL, Cristianini M (2014) Thermal processes—Commercial sterility (Retort). In: Batt CA (ed) Encyclopedia of food microbiology. Academic Press, Cambridge, pp 567–576. https://doi.org/10.1016/b978-0-12-384730-0.00405-5

Ball CO, Olson FCW (1957) Sterilization in food technology. Theory, practice, and calculations. McGraw-Hill Book Company, Inc., New York, Toronto, and London

Bennett RW, Berry MR Jr (1987) Serological reactivity and in vivo toxicity of *Staphylococcus aureus* enterotoxins A and D in selected canned foods. J Food Sci 52(2):416–419. https://doi.org/10.1111/j.1365-2621.1987.tb06628.x

Blackwell JH, Cliver DO, Callis JJ, Heidelbaugh ND, Larkin EP, McKERCHER PD, Thayer DW (1985) Foodborne viruses: their importance and need for research. J Food Prot 48(8):717–723. https://doi.org/10.4315/0362-028X-48.8.717

Blaschek HP (1999) Clostridium. *Clostridium Perfringens*. Encyclopedia of food microbiology, pp 433–438. https://doi.org/10.1006/rwfm.1999.0375

Brown KL (2000) Control of bacterial spores. Brit Med Bull 56(1):158–171. https://doi.org/10.1258/0007142001902860

Brooks EM, Houpt CW (1975) Canning process. US Patent 3,886,296 A, 27 May 1975

Carlin F (2011) Origin of bacterial spores contaminating foods. Food Microbiol 28(2):177–182. https://doi.org/10.1016/j.fm.2010.07.008

Carslaw HS, Jaeger JC (1959) Conduction of Heat in Solids. Clarendon Press, Oxford

Casolari A (1996a) I batteri sporigeni e la stabilizzazione degli alimenti. In: Ottaviani F (ed) Microbiologia dei Prodotti di Origine Vegetale – Ecologia ed Analisi Microbiologica. Chiriotti Editori, Pinerolo

Casolari A (1996b) Sterilizzazione: principi e applicazioni. In: Ottaviani F (ed) Microbiologia dei Prodotti di Origine Vegetale – Ecologia ed Analisi Microbiologica. Chiriotti Editori, Pinerolo

Charm SE, Landau SH (1987) Thermalizer. High-temperature short-time sterilization of heat-sensitive biological materials. Ann New York Acad Sci 506(1):608–612. https://doi.org/10.1111/j.1749-6632.1987.tb23855.x

Chen Z (2017) Microbial inactivation in foods by ultrasound. J Food: Microbiol Saf Hyg 02(1):1–2. https://doi.org/10.4172/2476-2059.1000e102

Chen Y, Yu LJ, Rupasinghe HV (2013) Effect of thermal and non-thermal pasteurisation on the microbial inactivation and phenolic degradation in fruit juice: a mini-review. J Sci Food Agric 93(5):981–986. https://doi.org/10.1002/jsfa.5989

Costa R (2003) Commercial food service establishments: the principles of modern food hygiene. In: Schmidt RH, Rodrick GE (eds) Food safety handbook, pp 453–522. Wiley, Hoboken, NJ, USA. https://doi.org/10.1002/047172159X.ch26

Couvert O, Gaillard S, Savy N, Mafart P, Leguérinel I (2005) Survival curves of heated bacterial spores: effect of environmental factors on Weibull parameters. Int J Food Microbiol 101(1):73–81. https://doi.org/10.1016/j.ijfoodmicro.2004.10.048

Dangerfield HG (1973) Effects of enterotoxins after ingestion by humans. In: Proceedings of the 73rd annual meeting of the American society for microbiology, vol 6, Miami Beach

Da Silva N, Taniwaki M, Junqueira V, De Arruda Silveira N, Da Silva Do Nascimento M, Gomes R (2013) Microbiological examination methods of food and water: a laboratory manual. CRC Press/Balkema, Taylor & Francis Group, pp 311–333. https://doi.org/10.1201/b13740-24

Dotzauer C, Ehrmann MA, Vogel RF (2002) Occurrence and detection of thermoanaerobacterium and thermoanaerobacter in canned food. Food Technol Biotechnol 40(1):21–26

Doyle ME, Mazzotta AS (2000) Review of studies on the thermal resistance of Salmonellae. J Food Prot 63(6):779–795. https://doi.org/10.4315/0362-028X-63.6.779

Engehnan MS, Sani RL (1983) Finite-element simulation of an inpackage pasteurization process. Numer Heat Transjkr 6(1):41–54. https://doi.org/10.1080/01495728308963073

Etoa FX, Michiels L (1988) Heat-induced resistance of *Bacillus stearothermophilus* spores. Lett Appl Microbiol 6(3):43–45. https://doi.org/10.1111/j.1472-765X.1988.tb01211.x

Evancho GM, Walls I (2001) Aciduric flat sour sporeformers. In: Compendium of methods for the microbiological examination of foods. American Public Health Association, Washington, DC, pp 239–244. https://doi.org/10.2105/9780875531755ch24

Fields ML (1970) The flat sour bacteria. Adv Food Res 18:163–217. https://doi.org/10.1016/s0065-2628(08)60370-5

Gates SD, Daniel McCartt A, Lappas P, Jeffries JB, Hanson RK, Hokama LA, Mortelmans KE (2010) Bacillus endospore resistance to gas dynamic heating. J Appl Microbiol 109(5):1591–1598. https://doi.org/10.1111/j.1365-2672.2010.04785.x

Gaze J (2005) Microbiological aspects of thermally processed foods. J Appl Microbiol 98 (6):1381–1386. https://doi.org/10.1111/j.1365-2672.2005.02636.x

Gómez-Sánchez A (2007) Microorganismos de importancia en el tratamiento térmico de alimentos ácidos y de alta acidez. Temas Selectos Ing Aliment 1:24–32

Gooch JW (2011) Flat sour spoilage. In: Good JW (ed) Encyclopedic dictionary of polymers. Springer. New York, pp 893–893. https://doi.org/10.1007/978-1-4419-6247-8_13764

Gupta R, Balasubramaniam VM (2012) High-pressure processing of fluid foods. In: Cullen PJ, Tiwari BK, Valdramidis VP (eds) Novel thermal and non-thermal technologies for fluid foods. Academic Press, London, Waltham, and San Diego, pp 109–133. https://doi.org/10.1016/b978-0-12-381470-8.00005-0

Hall RC (1971) Simple test to predict commercial sterility of heated food products. J Milk Food Technol 34(4):196–197. https://doi.org/10.4315/0022-2747-34.4.196

Heinz G, Hautzinger P (2007) Meat processing technology for small to medium scale producers. Food and Agriculture Organization of the United Nations, Regional office for Asia and the Pacific, Bangkok. ISBN: 978-974-7946-99-4. Available http://www.fao.org/3/a-ai407e.pdf. Accessed 17 Nov 2017

Henriques AO, Moran CP (2000) Structure and assembly of the bacterial endospore coat. Methods 20(1):95–110. https://doi.org/10.1006/meth.1999.0909

Hersom AC, Hulland ED (1981) Canned foods: thermal processing and microbiology, 7th edn. Chemical Publishing Company, New York

Icier F (2012) Ohmic heating of fluid foods. In: Cullen PJ, Tiwari BK, Valdramidis VP (eds) Novel thermal and non-thermal technologies for fluid foods. Academic Press, London, Waltham, and San Diego, pp 305–367. https://doi.org/10.1016/b978-0-12-381470-8.00011-6

Jay JM, Loessner MJ, Golden DA (eds) (2008a) Modern food microbiology, 7th edn. Springer Science & Business Media, New York, pp 727–745

Jay JM, Loessner MJ, Golden DA (eds) (2008b) Modern food microbiology, 7th edn. Springer Science & Business Media, New York, pp 301–350

Jay JM, Loessner MJ, Golden DA (eds) (2008c) Modern food microbiology, 7th edn. Springer Science & Business Media, New York, pp 13–37

Jay JM, Loessner MJ, Golden DA (eds) (2008d) Modern food microbiology, 7th edn. Springer Science & Business Media, New York, pp 567–590

Jay JM, Loessner MJ, Golden DA (eds) (2008e) Modern food microbiology, 7th edn. Springer Science & Business Media, New York, pp 415–441

Jay JM, Loessner MJ, Golden DA (eds) (2008f) Modern food microbiology, 7th edn. Springer Science & Business Media, New York, pp 709–726

Jen Y, Manson JE, Stumbo CR, Zahradnik JW (1971) A procedure for estimating sterilization of and quality factor degradation in thermally processed foods. J Food Sci 36(4):693–698. https://doi.org/10.1111/j.1365-2621.1971.tb15164.x

Johnson EA (1999) Clostridium—*Clostridium Botulinum*. Encyclopedia of food microbiology, pp 458–463. https://doi.org/10.1006/rwfm.1999.0395

Jongyingcharoen JS, Ahmad I (2014) Thermal and non-thermal processing of functional foods. In: Noomhorm A, Ahmad I, Anal AK (eds) Functional foods and dietary supplements: processing effects and health benefits. Wiley, Chichester, UK. https://doi.org/10.1002/9781118227800.ch11

Kihm DJ, Hutton MT, Hanlin JH, Johnson EA (1988) Zinc stimulates sporulation in *Clostridium botulinum* 113B. Curr Microbiol 17(4):193–198. https://doi.org/10.1007/BF01589451

Kitabatake N, Trivedi AB, Doi E (1991) Thermal decomposition and detoxification of citrinin under various moisture conditions. J Agric Food Chem 39(12):2240–2244. https://doi.org/10.1021/jf00012a028

Larousse J, Brown BE (eds) (1997) Food canning technology. Wiley-VCH Inc, New York, Chichester, Weinheim, Brisbane, Singapore, Toronto

Labbe RG, Duncan CL (1974) Sporulation and enterotoxin production by *Clostridium perfringens* type A under conditions of controlled pH and temperature. Can J Microbiol 20(11):1493–1501. https://doi.org/10.1139/m74-233

Lenz MK, Lund DB (1977) The lethality-Fourier number method: experimental verification of a model for calculating average quality factor retention in conduction-heated, canned foods. J Food Sci 42(4):989–996. https://doi.org/10.1111/j.1365-2621.1977.tb12652.x

Lund DB (1982) Applications of optimization in heat processing. Food Technol 36(7):97–100

Mañas P, Pagán R (2005) Microbial inactivation by new technologies of food preservation. J Appl Microbiol 98(6):1387–1399. https://doi.org/10.1111/j.1365-2672.2005.02561.x

Matsuda N, Komaki M, Ichikawa R, Gotoh S (1985) Aerobic and facultative anaerobic spore-forming bacteria isolated from spoiled canned foods. J Jpn Soc Food Sci Technol 32 (6):399–406. https://doi.org/10.3136/nskkk1962.32.6_399

Mauer LJ, Ozen BF (2004) Food Packaging. In: Scott Smith J, Hui YH (eds) Food processing: principles and applications. Blackwell Publishing, Ames and Oxford

Mazas M, Lopez M, Gonzalez I, Bernardo A, Martin R (1997) Effects of sporulation pH on the heat resistance and the sporulation of *Bacillus cereus*. Lett Appl Microbiol 25(5):331–334. https://doi.org/10.1046/j.1472-765X.1997.00240.x

McClane BA (2007) *Clostridium perfringens*. In: Doyle M. Beuchat L (eds) Food microbiology: fundamentals and frontiers, 3rd edn. ASM Press, Washington, DC, pp 423–444. https://doi.org/10.1128/9781555815912.ch19

McKenney PT, Driks A, Eichenberger P (2013) The *Bacillus subtilis* endospore: assembly and functions of the multilayered coat. Nat Rev Microbiol 11(1):33–44. https://doi.org/10.1038/nrmicro2921

Membré JM, van Zuijlen A (2011) A probabilistic approach to determine thermal process setting parameters: application for commercial sterility of products. Int J Food Microbiol 144(3):413–420. https://doi.org/10.1016/j.ijfoodmicro.2010.10.028

Mishkin M, Karel M, Saguy I (1982) Applications of optimization in food dehydration. Food Technol 36(7):10l–9

Miyamoto K, Nagahama M (2016) Clostridium: food poisoning by *Clostridium perfringens*. In: Caballero B, Finglas PM, Toldrá F (eds) Encyclopedia of food and health. Academic Press, Oxford, pp 149–154. https://doi.org/10.1016/b978-0-12-384947-2.00171-9

Moir A (2006) How do spores germinate? J Appl Microbiol 101(3):526–530. https://doi.org/10.1111/j.1365-2672.2006.02885.x

NFPA/CMI Container Integrity Task Force (1984) Botulism risk from post-processing contamination of commercially canned foods in metal containers. J Food Prot 47:801–816

Nightingale RW, Stallings D (1986) Assessing the extra-commercial food needs of low-income countries. Food Policy 11(1):27–41. https://doi.org/10.1016/0306-9192(86)90045-x

Notermans SHW (1999) Clostridium—Detection of neurotoxins of *Clostridium botulinum*. Encyclopedia of food microbiology, pp 463–466. https://doi.org/10.1006/rwfm.1999.0400

Olson KE, Sorrells KM (2015) 26. Thermophilic flat sour sporeformers. In: Salfinger Y, Tortorello ML (eds) Compendium of methods for the microbiological examination of foods. American Publih Health Association, Wagshington, DC. https://doi.org/10.2105/mbef.0222.031

Oomes SJCM, Van Zuijlen ACM, Hehenkamp JO, Witsenboer H, Van der Vossen JMBM, Brul S (2007) The characterisation of Bacillus spores occurring in the manufacturing of (low acid) canned products. Int J Food Microbiol 120(1):85–94. https://doi.org/10.1016/j.ijfoodmicro.2007.06.013

Paidhungat M, Setlow P (2002) Germination and outgrowth. In: Sonenshein AL, Losick R, Hoch JA (eds) *Bacillus subtilis* and its closest relatives: from genes to cells. American Society for Microbiology, pp 537–548. https://doi.org/10.1128/9781555817992.ch37

Peleg M (2003) Microbial survival curves: interpretation, mathematical modeling, and utilization. Comments Theor Biol 8(4–5):357–387. https://doi.org/10.1080/08948550302436

Peleg M, Cole MB (1998) Reinterpretation of microbial survival curves. Crit Rev Food Sci 38 (5):353–380. https://doi.org/10.1080/10408699891274246

Peleg M, Normand MD, Corradini MG (2005) Generating microbial survival curves during thermal processing in real time. J Appl Microbiol 98(2):406–417. https://doi.org/10.1111/j.1365-2672.2004.02487.x

Pflug IJ, Holcomb RG, Gómez MM (2001) Principles of the thermal destruction of microorganisms. In: Block SS (ed) Disinfection, sterilization, and preservation, 5th edn. Lippincott Williams & Wilkins, Philadelphia

Pflung IJ, Odlaugh TE (1978) A review of z and F values used to ensure the safety of low-acid canned foods. Food Technol 2:63–70

Rastogi NK (2012) Infrared heating of fluid foods. In: Cullen PJ, Tiwari BK, Valdramidis VP (eds) Novel thermal and non-thermal technologies for fluid foods. Academic Press, London, Waltham, and San Diego, pp 411–432. https://doi.org/10.1016/b978-0-12-381470-8.00013-x

Richards GP, McLeod C, Le Guyader FS (2010) Processing strategies to inactivate enteric viruses in shellfish. Food Environ Virol 2(3):183–193. https://doi.org/10.1007/s12560-010-9045-2

Richardson KC (1972) Microbial spoilage in Australian canned foods, 1955–68. Food Technol Aust 24:106–107

Sahin S, Sumnu SG (2006) Thermal properties of foods. In: Sahin S, Sumnu SG, Physical properties of foods, pp 107–155. https://doi.org/10.1007/0-387-30808-3_3

Schmitt HP (1966) Commercial sterility in canned foods, its meaning and determination. Assoc Food Drug Off US Q Bull 30:141–151

Setlow P, Johnson EA (2013) Spores and their significance. In: Doyle MP, Buchanan RL (eds) Food microbiology: fundamentals and frontiers, 4th edn. American Society of Microbiology, Washington, D.C., pp 45–79. https://doi.org/10.1128/9781555818463.ch3

Stumbo CR (1973) Thermobacteriology in food processing, 2nd edn. Academic Press Inc, New York

Teixeira AA (1971) Thermal process optimization through computer simulation of variable boundary control and container geometry. Dissertation, University of Massachusetts, Amherst

Tiwari BK, Mason TJ (2012) In: Cullen PJ, Tiwari BK, Valdramidis VP (eds) Novel Thermal and non-thermal technologies for fluid foods. Academic Press, London, Waltham, and San Diego, pp 135–165. https://doi.org/10.1016/b978-0-12-381470-8.00006-2

Tucker GS, Featherstone S (2011) Essentials of thermal processing. Wiley, New York

Vasseur C, Baverel L, Hebraud M, Labadie J (1999) Effect of osmotic, alkaline, acid or thermal stresses on the growth and inhibition of Listeria monocytogenes. J Appl Microbiol 86(3):469–476. https://doi.org/10.1046/j.1365-2672.1999.00686.x

Von Bockelmann BA, Von Bockelmann IL (1986) Aseptic packaging of liquid food products: a literature review. J Agric Food Chem 34(3):384–392. https://doi.org/10.1021/jf00069a001

Wells-Bennik MH, Eijlander RT, Den Besten HM, Berendsen EM, Warda AK, Krawczyk AO, Nierop Groot MN, Xiao Y, Zwietering MH, Kuipers OP, Abee T (2016) Bacterial spores in food: survival, emergence, and outgrowth. Ann Rev Food Sci Technol 7:457–482. https://doi.org/10.1146/annurev-food-041715-033144

Chapter 2
Failures of Thermally Treated Canned Foods

Angela Montanari, Caterina Barone, Michele Barone
and Anna Santangelo

Abstract The description of thermal treatments for canned foods is only the first step towards the comprehension of these products because of the implicit possibility of commercial and safety failures. This chapter discusses several of the most known defects on canned foods with relation to the entire product (food/packaging system), the food or the container only. The concept of 'failure' or 'defect' is the opposite concept of 'excellent' or 'good performance' that should be expected by canners. The 'right' approach should be the discussion of failures with reference to the global food/packaging system, the food only, or the container only. Failures of the whole system can be considered as the synergic effect of food- and packaging-related defects; there is abundant literature for these problems. On the contrary, literature concerning defects of the metal container—and related consequences—is not abundant, and this chapter aims to give more specific information to Readers.

Keywords Can coating · Canned food · Corrosion · Durability
Hermeticity · Mechanical resistance · Thermal treatment

Abbreviations

DFO   Department of Fisheries and Oceans
PVC   Polyvinyl chloride

2.1   Failures of Thermally Treated Canned Foods: An Introduction

The description of thermal treatments for canned foods (Chap. 1) is only the first step towards the comprehension of these products because of the implicit possibility of commercial and safety failures. Defects of canned foods can be related to the whole food/packaging system, to the edible content only or the 'accessory' packaging.

Thermal strategies have their 'weight' when speaking of defects. Thermal processes have the basic aim of destroying microorganisms (bacteria and

© The Author(s), under exclusive licence to Springer International Publishing AG,   17
part of Springer Nature 2018
A. Montanari et al., *Thermal Treatments of Canned Foods*,
Chemistry of Foods, https://doi.org/10.1007/978-3-319-74132-1_2

spore-forming life forms) in foods. The inhibition of microbial growth and the inactivation of microbial toxins are also needed. Consequently, the choice of the 'right' thermal treatment (pasteurisation and sterilisation) and related process parameters (time and temperature) have to be considered in different ambits, including canned foods (Chen 2017; Chen et al. 2013; Gupta and Balasubramaniam 2012; Icier 2012; Jongyingcharoen and Ahmad 2014; Mañas and Pagán 2005; Rastogi 2012; Sahin and Sumnu 2006; Tiwari and Mason 2012; Vasseur et al. 1999). For these reasons, this chapter shows and examines several of the most known defects on canned foods with relation to chemical failures (food, packaging or whole packaged system).

In general, thermal treatments are correlated with the concept of 'commercial sterility'. This idea should be considered with attention (Augusto et al. 2014; Codex Alimentarius Commission 1979; Costa 2003; Da Silva et al. 2013; Hall 1971; Heinz and Hautzinger 2007; Jay et al. 2008e; Membré and van Zuijlen 2011; Nightingale and Stallings 1986; Schmitt 1966), in relation to many food commodities, including canned foods. 'Commercially sterile' canned foods should remain stable for extended storage periods. On the other side, the term 'absolute sterility' does not appear to be applicable in all possible situations at present.

The discussion of failures related to canned foods packaged in steel-made metal containers can be interesting by the consumer's viewpoint because of the historical and traditional relationship between food preservation and food processing with the use of metallic containers. In other terms, there is a variegated group of thermally processed foods—vegetables, meat, fish, and fruits—which are generally correlated with the so-called 'tin can' because of traditional reasons and the same history of thermal processing (Brunazzi et al. 2014; Carle et al. 2001; Goldblith 1971; Joly 1996; Muys 1975).

The definition of 'food failure' or 'defect' should be made carefully because each food product in the modern world is—with important and restricted exceptions, depending on the market—a pre-packaged food (Parisi 2012, 2013). Moreover, canned foods are often processed foods because the simple packaging into metallic containers does not mean automatically 'food safety' in spite of the implicit and apparently accepted subliminal message for normal consumers (canned food = safe food). After all, double-seamed tin cans were named 'sanitary cans' (Pelley 1968; Stumbo 1973)!

Anyway, 'failure' or 'defect' is the opposite concept of 'excellent' or 'good performance', which should be expected by canners as the result of a reliable design, implementation and final processing itinerary (Parisi 2012). In other words, the performance of canned foods should be completely predictable because the original series of design, implementation and processing steps is composed of non-randomised steps. Each non-random subdivision of the whole process is correlated with previous and subsequent steps in a logical and reliable succession: the final result should be the logical consequence (Parisi 2005).

On the other hand, process deviations from the 'ideality'—the predictable result—should be investigated on the basis of process failures. In other words, there not should exist unintelligible defects in a completely monitored process.

This premise is needed because the definition of 'canned food failures' has to take into account all possible and predictable—or unpredictable—defects of the whole food/packaging system (the food product): these defects represent all deviations from the ideality, of the contrary of the expected result in terms of positive properties (Parisi 2004).

Consequently, positive features of canned foods should be always declared (Fig. 2.1) before speaking of negative results. Canned foods are generally expected to offer (Brunazzi et al. 2014):

(a) Long durability, depending on food preservation treatments before, during and after packaging;
(b) No necessity of low thermal values during storage, with possible exceptions;
(c) Excellent or good mechanical resistance against logistic-related damages during pasteurisation or sterilisation treatments;
(d) Reduced risks caused by air oxidation because of barrier properties and little air amounts into canned foods;
(e) Excellent protection against light exposure and
(f) Possible use of containers as self-cooling or self-heating cans (actually, this option seems sporadic enough).

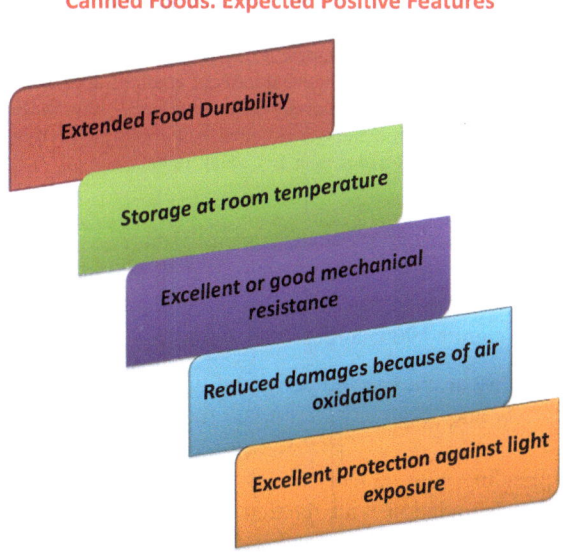

Canned Foods. Expected Positive Features

Extended Food Durability

Storage at room temperature

Excellent or good mechanical resistance

Reduced damages because of air oxidation

Excellent protection against light exposure

Fig. 2.1 Positive features of canned foods. In general, these products are expected to offer extended shelf-life performances, depending on food preservation treatments before, during and after packaging. In addition, easy storage conditions are possible (room temperature). Logistic-related damages are quite limited; excellent barrier properties, protection against light exposure, and hermetic closures can give additional food safety advantages. The use of related containers as self-cooling or self-heating cans does not seem one of most common applications at present

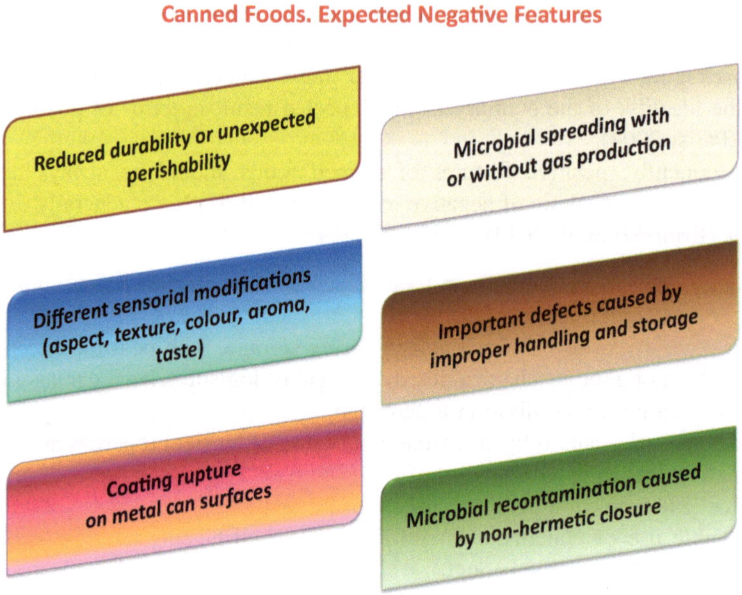

Canned Foods. Expected Negative Features

Reduced durability or unexpected perishability

Microbial spreading with or without gas production

Different sensorial modifications (aspect, texture, colour, aroma, taste)

Important defects caused by improper handling and storage

Coating rupture on metal can surfaces

Microbial recontamination caused by non-hermetic closure

Fig. 2.2 Predictable negative performances of canned foods. By the hygienic viewpoint, decreased shelf-life features or unexpected perishability of long-durability canned foods could be expected. Microbial recontamination caused by non-hermetic closure has been reported. At the same time, canned food might show microbial spreading with or without gas production, organoleptic degradation with concern to apparent density, texture, colour, aroma and taste. The improper handling and storage may cause notable defects; coating ruptures on the inner and the external sides of the container may be signalled. Other situations—rare food oxidation episodes with or without gas production, food safety risks after self-cooling or self-heating processes— might be considered, but their importance does not seem high

As a result, negative performances of canned foods (Fig. 2.2) should be at least (Dotzauer et al. 2002; Gómez-Sánchez 2007; Jay et al. 2008a, b, c, d; Matsuda et al. 1985; NFPA/CMI Container Integrity Task Force 1984; Pflug et al. 1981; Richardson 1972; Schmitt 1966; Setlow and Johnson 2013):

(1) Reduced durability or unexpected perishability of long-durability canned foods. This is obviously one of the most important food safety dangers by the hygienic viewpoint;

(2) Microbial spreading with or without gas production (Sect. 1.2.3);

(3) Sensorial food degradation with relation to aspect (apparent density), texture (where applicable), colour, aroma and taste. The increase of acidity values may be observed in certain situations;

(4) Important defects of the whole system (food + container), or the container only, or the contained food, caused by improper handling and storage;

(5) Coating rupture with superficial damages (on the inner and the external sides of the container), with the migration of metal or of organic fragments (Parkar and Rakesh 2014; Piergiovanni and Limbo 2016);
(6) Rare food oxidation with or without gas production due to non-hermetic closure in high-vacuum metal cans;
(7) Microbial recontamination caused by non-hermetic closure and
(8) Food safety risks after self-cooling or self-heating processes (this situation is quite sporadic).

This list of failures is not exhaustive; in addition, some of above-mentioned defects could have more than one simple cause. For these reasons, these failures should be described with relation to the passive subject of alteration: the whole system (food/packaging), the contained food only or the metal container only.

2.2   Thermally Treated Canned Foods: General Failures

By the hygienic viewpoint, foods have to be guaranteed safe, legally compliant and correspondent to their description (otherwise, authenticity and economically motivated adulteration issues could occur). Because of the interconnected relationship between contained food and metallic container in canned foods—a two-binaries interchange system where food could be modified and modify the container and vice versa—each food safety risk should be potentially ascribed in principle to the whole synergic agglomeration (Brunazzi et al. 2014; Parisi 2012, 2013).

In this ambit, the following defects with hygienic meaning may be listed as follows (Brunazzi et al. 2014; Montanari 2015a):

(a) Detection of foreign metal ions in the food: iron and tin;
(b) Sporadic detection of foreign organic compounds in the food; visual appearance of foreign and incoherent coating materials into the food;
(c) Suspect damages on container inner sides, including strange microbubbles and
(d) Apparent damages on container external sides, including graphic appearance (brands, logo, etc.).

The above-shown list (Fig. 2.3) may be not completely exhaustive, although many of the commonly observed defects are mentioned. These failures have a peculiar common point: one or more important deviations from the ideality with consequent safety consequences. On the other hand, certain non-food safety-related failures might be observed.

In general, defects of the entire and synergic food/packaging systems may be ascribed to one or more of the following causes (AACC International 2009; Blaschek 1999; Brunazzi et al. 2014; Evancho and Walls 2001; Fields 1970; Gooch 2011; Johnson 1999; McClane 2007; Miyamoto and Nagahama 2016; Montanari 2015b; Notermans 1999; Olson and Sorrells 2015; Parisi 2012, 2013):

Fig. 2.3 Main failures of the food/packaging system in thermally treated canned foods. In general, these defects concern the detection of tin and iron and the rare detection of foreign and incoherent coating materials into the food. In addition, container inner sides might show strange superficial damages; other defects might be observed on external sides

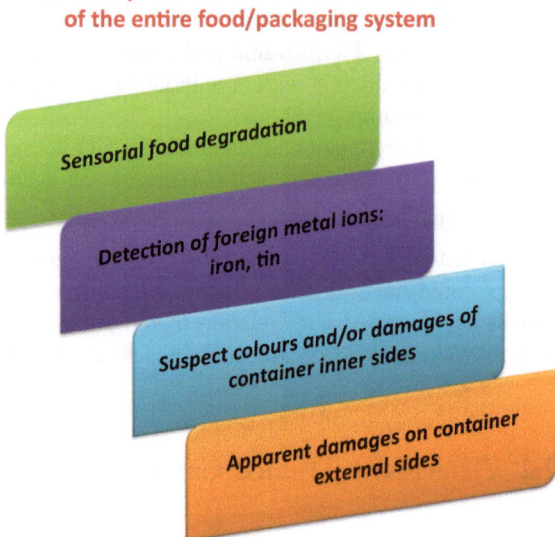

(1) Degradation of the edible content because of (a) inadequate quality features from the microbiological viewpoint and improper thermal treatment at the same time (Chap. 1), and (b) post-process contamination (spoilage) in low-acid canned foods because of incorrect 'good manufacturing practices'. Good manufacturing practices should concern at least: the use of cooling water with good microbiological quality (sufficient disinfection); limitation of container handling; disinfection of packaging handling instruments; proper segregation with minimisation of cross-contamination episodes (Anonymous 1969; Larousse and Brown 1997);

(2) Storage-related defects, including degradation of the edible content, for example, altered colours, strange off-colours, etc., because of improper chemical and physical features, including excessive acidity and microbial contamination. Interestingly, advanced microbial spreading may partially destroy and degrade the food before thermal processing and packaging, with predictable and irreversible results (adequate thermal treatments might worse sensorial features of the food). It should be also noted that the migration of metal ions such as iron (Page et al. 2003) into foods might be detrimental. This metal can be a good catalysing agent for certain degradative reactions, and its action depends also on other environmental conditions such as storage temperatures. In addition, excessive light exposure could 'heat up' canned foods with possible thermal increase and consequent degradation phenomena into metal cans (Anonymous 1969; Montanari 2015b) and

(3) Corrosion-related defects of the metallic support and/or the final container, for example, superficial scratches with tin removal from the surface insufficient coating, rheological problems concerning the coating process (excessive viscosity values may cause the irregular deposition of the lacquer on metal surfaces), mechanical damages (coating removal) including superficial scratches, corrosion because of insufficient coating protection, ink 'bleeding', insufficient coating adhesion, etc.

Because of the complexity of this matter, the 'right' approach should be the discussion of failures depending on the main cause or causes with relation to the global food/packaging system, the food only or the container only. However, the scientific literature contains many studies concerning the stability—and potential defects—of canned foods depending on microbiological chemical–physical, and processing features of the complete process (including thermal treatments and related failures). The analytical approach should be considered when speaking of visually appreciable defects: in this ambit, technologies such as X-rays inspection and digital image and processing evaluations may be potentially helpful (Correia and Mittal 1986; De Sitter and van de Haar 1998; Du and Sun 2004; Hui 2007; Leon et al. 2006; Maijala 2014; Mendoza et al. 2006; Parmee 1990; Upmann and Bonaparte 1999; Yam and Papadakis 2004).

Chapters 1 and 4 are dedicated to the degradation of canned foods caused by food-related and processing-related factors; on the other hand, failures of the whole system may be considered as the synergic effect of food- and packaging-related defects. For this reason, the following section gives a simple description of container-related defects with some link to commonly observed failures of canned foods. In addition, a similar discussion should take into account a certain knowledge of metal can technology; consequently, the Reader is invited to consult more specific texts with relation to this peculiar industrial sector.

2.3 Thermally Treated Canned Foods: Container-Related Defects

In general, container-related failures of canned foods depend on the polymeric nature of the coated metal surface because of the essential metallic structure (tinplate or tin-free steel materials for pasteurised or sterilised canned foods) on the one side, and the necessary protection of supports, depending on the required food application (Larousse and Brown 1997; Parisi 2004, 2012).

Therefore, the identification of most common metal can failures is not apparently difficult and it can be expressed in a simplified way as follows. Some of these defects, related to insufficient coating quantities on metal surfaces, can be briefly described as an example. These defects do not correspond to the entire group of currently known failures, but their importance is normally recognised (and related damages are enhanced) after sterilisation or pasteurisation treatments (Fig. 2.4).

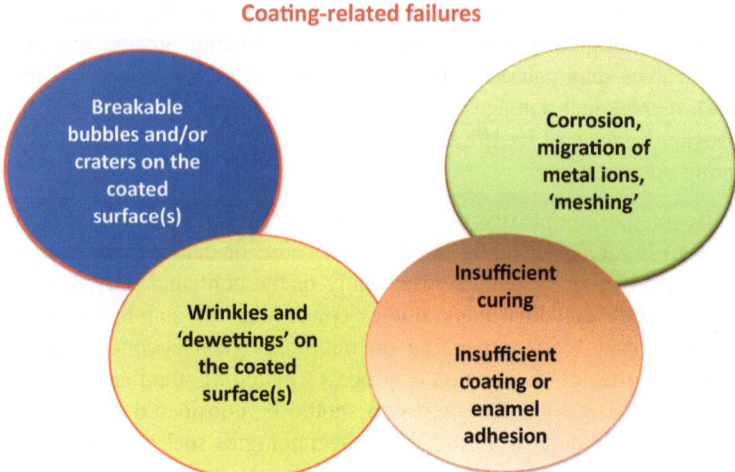

Fig. 2.4 Coating-related failures in thermally treated canned foods. The coating process is technically a curing process of deposed resins on metal surfaces. Should this step be incorrectly performed, different superficial defects could be observed on the inner and the external side of metal cans, including breakable bubbles and/or craters, 'dewetting' (many pinholes without coating), insufficient curing and/or adhesion of coatings or enamels to the metallic support, etc.

2.3.1 Insufficient Coating Amount

Tinplate and tin-free steel alloy materials have to be coated with liquid paints able to give higher resistance against environmental agents and food or beverage products (Piergiovanni and Limbo 2016). This resistance is especially required on the inner side of metal cans, while the outer wall has to receive generally lithographic pictures; for this reason and for protection exigencies also, both sides of original tinplate sheets have to be coated (Pilley 1981; Oldring and Nehring 2007) with adequate lacquers. The coating process is technically a curing process of deposed resins on metal surfaces; a transparent (finishing) coating may be required on the inner and the outer sides with the aim of protecting pictures and side-seam junctions. This area is very critical and the protection is also needed on the inner side (Parisi 2004).

Should this critical step—the coating process—be incorrectly performed, the following failures would be observed with extremely important consequences (Montanari 2015a):

(a) Appearance of breakable bubbles and/or craters on the coated surface(s);
(b) Possibility of wrinkles and 'dewettings' or eyeholings (many pinholes without coating, diffused on the entire sheet);
(c) Insufficient curing with 'living' polymers, able to react with foods and beverages;

(d) Insufficient coating or enamel adhesion;
(e) Migration of metal ions (tin, iron, chromium, etc.), corrosion and incorporation (meshing) of organic molecules from the food to the plastic network;
(f) Defects of the superimposed coating structure. Should an enamel be chosen on the outer side, a transparent coating would be used as protection for superimposed inks (in a lithographic system). However, the adhesion of each product could be compromised in the coating step and
(g) Sulphur staining failures (these defects have aesthetical importance only).

References

AACC International (2009) Thermophilic Spore Counts (Total Aerobic, Flat-Sour, H$_2$S, Non-H$_2$S Anaerobic). Approved Methods of Analysis, 11th edn. Method 42-40-01. American Association of Cereal Chemists (AACC) International, St. Paul. https://doi.org/10.1094/aaccintmethod-42-40.01

Anonymous (1969) How to obtain best service form food cans, 3rd edn. Can Manufacturers Institute, Washington DC

Augusto PED, Tribst AAL, Cristianini M (2014) Thermal processes—commercial sterility (retort). In: Batt CA (ed) Encyclopedia of food microbiology. Academic Press, Cambridge, pp 567–576. https://doi.org/10.1016/b978-0-12-384730-0.00405-5

Blaschek HP (1999) Clostridium. *Clostridium perfringens*. Encyclopedia of food microbiology, pp 433–438. https://doi.org/10.1006/rwfm.1999.0375

Brunazzi G, Parisi S, Pereno A (2014) Packaging and food: a complex combination. In: Brunazzi G, Parisi S, Pereno A (eds) The importance of packaging design for the chemistry of food products. SpringerBriefs in chemistry of foods. Springer International Publishing, Cham. https://doi.org/10.1007/978-3-319-08452-7_2

Carle R, Borzych P, Dubb P, Siliha H, Maier O (2001) A new process for firmer canned cherries and strawberries. Food Aust 53(8):343–348

Chen Z (2017) Microbial inactivation in foods by ultrasound. J Food Microbiol Saf Hyg 02(1):1–2. https://doi.org/10.4172/2476-2059.1000e102

Chen Y, Yu LJ, Rupasinghe HV (2013) Effect of thermal and non-thermal pasteurisation on the microbial inactivation and phenolic degradation in fruit juice: a mini-review. J Sci Food Agric 93(5):981–986. https://doi.org/10.1002/jsfa.5989

Codex Alimentarius Commission (1979) Code of hygienic practice for low and acidified low acid canned foods, CAC/RCP 23-1979. The Codex Alimentarius Commission, The Food and Agriculture Organization of the United Nations, Rome

Correia LR, Mittal GS (1986) Application of computer-aided digital image analysis in food processing. Can Inst Food Sci Technol J 19(4):xlvi. https://doi.org/10.1016/s0315-5463(86)71600-3

Costa R (2003) Commercial food service establishments: the principles of modern food hygiene. In: Schmidt RH, Rodrick GE (eds) Food safety handbook. Wiley, Hoboken, pp 453–522. https://doi.org/10.1002/047172159x.ch26

Da Silva N, Taniwaki M, Junqueira V, De Arruda Silveira N, Da Silva Do Nascimento M, Gomes R (2013) Microbiological examination methods of food and water: a laboratory manual. CRC Press/Balkema, Taylor & Francis Group, pp 311–333. https://doi.org/10.1201/b13740-24

De Sitter H, van de Haar S (1998) Governmental food inspection and HACCP. Food Control 9(2–3):131–135. https://doi.org/10.1016/s0956-7135(98)00083-8

Dotzauer C, Ehrmann MA, Vogel RF (2002) Occurrence and detection of thermoanaerobacterium and thermoanaerobacter in canned food. Food Technol Biotechnol 40(1):21–26

Du CJ, Sun DW (2004) Recent developments in the applications of image processing techniques for food quality evaluation. Trends Food Sci Technol 15(5):230–249. https://doi.org/10.1016/j. tifs.2003.10.006

Evancho GM, Walls I (2001) Aciduric flat sour sporeformers. Compendium of methods for the microbiological examination of foods. American Public Health Association, Washington, DC, pp 239–244. https://doi.org/10.2105/9780875531755ch24

Fields ML (1970) The flat sour bacteria. Adv Food Res 18:163–217. https://doi.org/10.1016/ s0065-2628(08)60370-5

Goldblith SA (1971) Thermal processing of foods a review. In: Bourne GH (ed) World review of nutrition and dietetics, vol. 13. Karger Publishers, Basel. https://doi.org/10.1159/000391885

Gómez-Sánchez A (2007) Microorganismos de importancia en el tratamiento térmico de alimentos ácidos y de alta acidez. Temas Selectos Ing Aliment 1:24–32

Gooch JW (2011) Flat sour spoilage. In: Good JW (ed) Encyclopedic dictionary of polymers, pp 893–893. Springer, New York. https://doi.org/10.1007/978-1-4419-6247-8_13764

Gupta R, Balasubramaniam VM (2012) High-pressure processing of fluid foods. In: Cullen PJ, Tiwari BK, Valdramidis VP (eds) Novel thermal and non-thermal technologies for fluid foods. Academic Press, London, Waltham, and San Diego, pp 109–133. https://doi.org/10.1016/b978-0-12-381470-8.00005-0

Hall RC (1971) Simple test to predict commercial sterility of heated food products. J Milk Food Technol 34(4):196–197. https://doi.org/10.4315/0022-2747-34.4.196

Heinz G, Hautzinger P (2007) Meat processing technology for small to medium scale producers. Food and agriculture organization of the United Nations, regional office for Asia and the Pacific, Bangkok. ISBN: 978-974-7946-99-4. Available http://www.fao.org/3/a-ai407e.pdf. Accessed 17 Nov 2017

Hui YH (2007) Food establishment inspection. In: Hui YH, Chandan RC, Clark S, Cross NA, Dobbs JC, Hurst WJ, Nollet LML, Shimoni E, Sinha N, Smith EB, Surapat S, Toldrá F, Titchenal A (eds) Handbook of food products manufacturing, Wiley, Hoboken. https://doi.org/ 10.1002/9780470113554.ch13

Icier F (2012) Ohmic heating of fluid foods. In: Cullen PJ, Tiwari BK, Valdramidis VP (eds) Novel thermal and non-thermal technologies for fluid foods. Academic Press, London, Waltham, and San Diego, pp 305–367. https://doi.org/10.1016/b978-0-12-381470-8.00011-6

Jay JM, Loessner MJ, Golden DA (eds) (2008a) Modern food microbiology, 7th edn. Springer Science & Business Media, New York, pp 13–37

Jay JM, Loessner MJ, Golden DA (eds) (2008b) Modern food microbiology, 7th edn. Springer Science & Business Media, New York, pp 301–350

Jay JM, Loessner MJ, Golden DA (eds) (2008c) Modern food microbiology, 7th edn. Springer Science & Business Media, New York, pp 415–441

Jay JM, Loessner MJ, Golden DA (eds) (2008d) Modern food microbiology, 7th edn. Springer Science & Business Media, New York, pp 567–590

Jay JM, Loessner MJ, Golden DA (eds) (2008e) Modern food microbiology, 7th edn. Springer Science & Business Media, New York, pp 415–441

Johnson EA (1999) Clostridium—*Clostridium botulinum*. Encyclopedia of food microbiology, pp 458–463. https://doi.org/10.1006/rwfm.1999.0395

Joly X (1996) L'industrie agro-alimentaire et l'environnement: Retour d'expérience sur le secteur de l'appertise. Revue de cytologie et de biologie végétales, Le Botaniste 19(3–4):61–63

Jongyingcharoen JS, Ahmad I (2014) Thermal and non-thermal processing of functional foods. In: Noomhorm A, Ahmad I, Anal AK (eds) Functional foods and dietary supplements: processing effects and health benefits. Wiley, Chichester, UK. https://doi.org/10.1002/9781118227800. ch11

Larousse J, Brown BE (eds) (1997) Food canning technology. Wiley-VCH Inc., New York, Chichester, Weinheim, Brisbane, Singapore, Toronto

Leon K, Mery D, Pedreschi F, Leon J (2006) Color measurement in L∗ a∗ b∗ units from RGB digital images. Food Res Int 39(10):1084–1091. https://doi.org/10.1016/j.foodres.2006.03.006

Maijala R (2014) Scientific risk assessment-basis for food legislation. In: Ninios T, Lundén J, Korkeala H, Fredriksson-Ahomaa M (eds) Meat inspection and control in the slaughterhouse. Wiley, Chichester. https://doi.org/10.1002/9781118525821.ch26

Mañas P, Pagán R (2005) Microbial inactivation by new technologies of food preservation. J Appl Microbiol 98(6):1387–1399. https://doi.org/10.1111/j.1365-2672.2005.02561.x

Matsuda N, Komaki M, Ichikawa R, Gotoh S (1985) Aerobic and facultative anaerobic spore-forming bacteria isolated from spoiled canned foods. J Jpn Soc Food Sci Technol 32(6):399–406. https://doi.org/10.3136/nskkk1962.32.6_399

McClane BA (2007) Clostridium perfringens. In: Doyle M. Beuchat L (eds) Food microbiology: fundamentals and frontiers, 3rd edn. ASM Press, Washington, DC, pp 423–444. https://doi.org/10.1128/9781555815912.ch19

Membré JM, van Zuijlen A (2011) A probabilistic approach to determine thermal process setting parameters: application for commercial sterility of products. Int J Food Microbiol 144(3):413–420. https://doi.org/10.1016/j.ijfoodmicro.2010.10.028

Mendoza F, Dejmek P, Aguilera JM (2006) Calibrated color measurements of agricultural foods using image analysis. Postharvest Biol Technol 41(3):285–295. https://doi.org/10.1016/j.postharvbio.2006.04.004

Miyamoto K, Nagahama M (2016) Clostridium: food poisoning by Clostridium perfringens. In: Caballero B, Finglas PM, Toldrá F (eds) Encyclopedia of food and health. Academic Press, Oxford, pp 149–154. https://doi.org/10.1016/b978-0-12-384947-2.00171-9

Montanari (2015a) Inorganic contaminants of food as a function of packaging features. In: Barone C, Bolzoni L, Caruso G, Montanari A, Parisi S, Steinka I (2015) Food packaging hygiene. SpringerBriefs in chemistry of foods. Springer International Publishing, Cham. https://doi.org/10.1007/978-3-319-14827-4_2

Montanari (2015b) Basic principles of corrosion of food metal packaging. In: Barone C, Bolzoni L, Caruso G, Montanari A, Parisi S, Steinka I (2015) Food packaging hygiene. SpringerBriefs in chemistry of foods. Springer International Publishing, Cham. https://doi.org/10.1007/978-3-319-14827-4_6

Muys GT (1975) Microbial safety and stability of food products. Antonie Van Leeuwenhoek 41 (1):369–371. https://doi.org/10.1007/BF02565074

NFPA/CMI Container Integrity Task Force (1984) Botulism risk from post-processing contamination of commercially canned foods in metal containers. J Food Prot 47:801–816

Nightingale RW, Stallings D (1986) Assessing the extra-commercial food needs of low-income countries. Food Policy 11(1):27–41. https://doi.org/10.1016/0306-9192(86)90045-x

Notermans SHW (1999) Clostridium—detection of neurotoxins of Clostridium botulinum. Encyclopedia of food microbiology, pp 463–466. https://doi.org/10.1006/rwfm.1999.0400

Oldring PKT, Nehring U (2007) Packaging materials 7: metal packaging for foodstuffs. International Life Science Institute Europe, Brussels

Olson KE, Sorrells KM (2015). Thermophilic flat sour sporeformers. In: Salfinger Y, Tortorello ML (eds) Compendium of methods for the microbiological examination of foods. American Public Health Association, Washington, DC. https://doi.org/10.2105/mbef.0222.031

Page B, Edwards M, May N (2003) Metal cans. In: Coles R, McDowell D, Kirwan MJ (eds) Food Packag Technol. Blackwell Publishing Ltd., Oxford

Parisi S (2004) Alterazioni in imballaggi metallici termicamente processati. Gulotta Press, Palermo

Parisi S (2005) La produzione "continua" è anche "costante"? Confutazione di alcuni luoghi comuni nel settore industriale/manifatturiero. Il Chimico Italiano 16(3–4):10–18

Parisi S (2012) Food packaging and food alterations: the user-oriented approach. Smithers

Parisi S (2013) Food industry and packaging materials performance-oriented guidelines for users. Smithers Rapra Technology, Shawbury

Parkar J, Rakesh M (2014) Leaching of elements from packaging material into canned foods marketed in India. Food Control 40:177–184. https://doi.org/10.1016/j.foodcont.2013.11.042

Parmee R (1990) X-rays give in-depth inspection. Sens Rev 10(2):84–86. https://doi.org/10.1108/eb007818

Pelley DL (1968) Combined sanitary can top cover and can tab opening hook. US Patent 3,362,572 A, 9 Jan 1986

Pflug IJ, Davidson PM, Holcomb RG (1981) Incidence of canned food spoilage at the retail level. J Food Prot 44(9):682–685. https://doi.org/10.4315/0362-028X-44.9.682

Piergiovanni L, Limbo S (2016) Food packaging materials. SpringerBriefs in chemistry of foods. Springer International Publishing, Cham. https://doi.org/10.1007/978-3-319-24732-8

Pilley KP (1981) Lacquers, varnishes and coatings for food and drink cans and for the decorating industry. Arthur Holden Surface Coatings Ltd., Birmingham

Rastogi NK (2012) Infrared heating of fluid foods. In: Cullen PJ, Tiwari BK, Valdramidis VP (eds) Novel thermal and non-thermal technologies for fluid foods. Academic Press, London, Waltham, and San Diego, pp 411–432. https://doi.org/10.1016/b978-0-12-381470-8.00013-x

Richardson KC (1972) Microbial spoilage in Australian canned foods, 1955–68. Food Technol Aust 24:106–107

Sahin S, Sumnu SG (2006) Thermal properties of foods. In: Sahin S, Sumnu SG (eds) Physical properties of foods, pp 107–155. https://doi.org/10.1007/0-387-30808-3_3

Schmitt HP (1966) Commercial sterility in canned foods, its meaning and determination. Assoc Food Drug Off US Q Bull 30:141–151

Setlow P, Johnson EA (2013) Spores and their significance. In: Doyle MP, Buchanan RL (eds) Food microbiology: fundamentals and frontiers, 4th edn. American Society of Microbiology, Washington, DC, pp 45–79. https://doi.org/10.1128/9781555818463.ch3

Stumbo CR (1973) Thermobacteriology in food processing, 2nd edn. Academic Press Inc, New York

Tiwari BK, Mason TJ (2012) In: Cullen PJ, Tiwari BK, Valdramidis VP (eds) Novel thermal and non-thermal technologies for fluid foods. Academic Press, London, Waltham, and San Diego, pp 135–165. https://doi.org/10.1016/b978-0-12-381470-8.00006-2

Upmann M, Bonaparte C (1999) Rapid methods for food hygiene inspection. Encyclopedia of food microbiology, pp 1887–1895. https://doi.org/10.1006/rwfm.1999.1320

Vasseur C, Baverel L, Hebraud M, Labadie J (1999) Effect of osmotic, alkaline, acid or thermal stresses on the growth and inhibition of *Listeria monocytogenes*. J Appl Microbiol 86(3):469–476. https://doi.org/10.1046/j.1365-2672.1999.00686.x

Yam KL, Papadakis SE (2004) A simple digital imaging method for measuring and analyzing color of food surfaces. J Food Eng 61(1):137–142. https://doi.org/10.1016/s0260-8774(03)00195-x

Chapter 3
Metal Cans and Canned Foods: Image Analysis of Visual Failures

**Angela Montanari, Caterina Barone, Michele Barone
and Anna Santangelo**

Abstract The qualitative examination of canned foods can be performed with many possible options, depending on the desired result. The microbiological evaluation of canned foods requires generally microbial examination testing methods. In addition, the chemical risk has to be evaluated in general with relation to the possible detection of undeclared allergens, genetically modified organisms, mycotoxins, pesticide residues, etc. Finally, the evaluation of food and beverage products can be carried out by means of sensorial testing methods. In this ambit, simple colorimetric tests may be created and implemented for industrial quality control purpose, and some of these procedures are direct expression of 'digital image analysis and processing' systems. Consequently, the possible alteration of certain tints can be analysed and critically discussed on condition that a reliable relationship has been established between the above-mentioned chromatic modification and the cause. This chapter is dedicated to 'digital image analysis and processing' practical applications for the evaluation of thermally treated canned foods.

Keywords Canned food · Colorimetry · Digital image analysis and processing
Light intensity · Pigment · Pixel · Software

Abbreviations

ALI	Average light intensity
DIAP	Digital image analysis and processing
ISO	International Organization for Standardization
LLI	Lycopene loss index
RGB	Red/green/blue

29
A. Montanari et al., *Thermal Treatments of Canned Foods*,
Chemistry of Foods, https://doi.org/10.1007/978-3-319-74132-1_3

3.1 Metal Cans and Canned Foods: The Image Analysis Strategy

The qualitative examination of canned foods can be performed with many possible options, depending on the desired result (Parisi 2013).

The microbiological evaluation of canned foods requires generally microbial examination testing methods: the identification of certain 'microbial markers' has to be assured before choosing the 'right' approach, and the same thing can be affirmed when speaking of pathogens (Andreis and Ottaviani 2002; Downes and Ito 2001).

The choice of 'target' microorganisms does not imply necessarily food degradation or specific direct effects on human safety (examples: *Pseudomonas* spp., Lactobacillaceae, yeasts and moulds, etc.); however, their presence may be used as good indicators of the hygienic processing in food industries, depending on products (Delia et al. 2005; Ottaviani 1996). With relation to pathogen agents, the choice is surely simple: this group is restricted to the following names at least (AACC International 2009; Blaschek 1999; Dotzauer et al. 2002; Evancho and Walls 2001; Fields 1970; Gooch 2011; Jay et al. 2008a; Johnson 1999; Matsuda et al. 1985; McClane 2007; Miyamoto and Nagahama 2016; Notermans 1999; Olson and Sorrells 2015; Pflug et al. 1981; Richardson 1972):

1. *Bacillus cereus,*
2. *Clostridium botulinum,*
3. *Clostridium perfringens,*
4. *Escherichia coli,*
5. *Listeria monocytogenes,*
6. *Salmonella* spp.,
7. *Staphylococcus aureus* and
8. *Campylobacter jejuni.*

However, the above-mentioned group is not completely exhaustive, and the presence (or the absence) of peculiar pathogens has to be justified with reference to the specific food and detailed reasons (Andreis and Ottaviani 2002; Delia et al. 2005) such as the demonstration of favourable or unfavourable factors (examples: water activity, pH values, redox potential values, etc.).

The chemical risk has to be evaluated in general with relation to the possible detection of (Bhagat et al. 2016; de Fátima Pocas and Hogg 2007; Ikem and Egiebor 2005; Jay et al. 2008b; Parisi 2016a):

(1) Undeclared allergens and food additives: This factor has to be specifically considered as part of the control of food labels, in accordance with the most recent norms such as the Food Safety Modernisation Act;
(2) Genetically modified organisms;
(3) Mycotoxins;

(4) Pesticide residues, dioxins, other non-food related chemical compounds with safety concerns and

(5) Other contaminants: heavy metals, radioactive substances, veterinary medical residues.

The above-mentioned list should be considered as the evolution of current regulatory norms (Pisanello 2014); for this reason, many possible dangers could be absent. On the other hand, quality controls concern also normal chemical and physical features of canned foods; consequently, this discussion should concern the sector of analytical chemistry with all possible methods, from old and traditional procedures to the most recent analytical instrumental techniques (Parisi 2016b).

Finally, the evaluation of food and beverage products can be also carried out by means of sensorial testing methods with the aim of determining indirectly the acceptability or unacceptability in a simplified and reliable way. Obviously, a simple organoleptic test evaluates an easily recognisable feature of the food or beverage product on condition that a specific relationship exists between this property and one (or more) important microbiological, chemical or physical parameters (Delia et al. 2005; Oliveira et al. 2009). The estimation of durability values may be a useful option; certain 'smart packaging' systems are explicitly dedicated to express residual freshness in a reliable way (Parisi 2009; Piergiovanni and Limbo 2016). Because of the influence of thermal strategies on certain failures, sensorial testing method may be very interesting in many industrial areas. The choice of the 'right' thermal treatment (pasteurisation, sterilisation) and related process parameters (time, temperature) in many industrial applications (Chen 2017; Chen et al. 2013; Gupta and Balasubramaniam 2012; Icier 2012; Jongyingcharoen and Ahmad 2014; Mañas and Pagán 2005; Rastogi 2012; Sahin and Sumnu 2006; Tiwari and Mason 2012; Vasseur et al. 1999) may be carried out on the basis of organoleptic results.

The examination of canned food concerns also metal cans (Chap. 2). In detail, food containers have to guarantee—in connection with foods or beverages, and adequate procedures—the final and expected result: a completely safe, legal and reliable food product (Parisi 2004–2012; Savov and Kouzmanov 2009). This result may be obtained with the risk assessment strategy and the creation/implementation of good manufacturing practices based on prerequisite programmes, as also required by the document ISO/TS 22002-4 (formerly PAS 223) concerning prerequisite programmes and design requirements for food safety in the manufacture and provision of food packaging (ISO 2013). On these bases, important requirements such as packaging 'workability'[1] and 'technological suitability'[2] can be defined and correctly evaluated (Bordonaro 2012).

---

[1]Workability means the effective performance of food packaging materials during and after their use into packing and processing lines.

[2]Technological suitability or 'performance' of a peculiar food packaging material is the ability to comply with specific norms and mandatory regulations on the one hand, and with the 'intended use(s)' of the material on the other side.

An additional approach to the problem of quality testing methods may be offered by 'hybrid' procedures: these systems are the multidisciplinary answer of food and food packaging technologists to particular situations where chemistry, microbiology, science of materials and other disciplines may be interconnected (Parisi 2013).

In this ambit, there are different analytical options with concern to the qualitative and quantitative examination of metal cans and packaged foods, including the estimation of certain 'visual' (explicit) failures. For this reason, simple colorimetric tests may be created and implemented for industrial quality control purpose, and some of these procedures are direct expression of 'digital image analysis and processing' (DIAP) systems.

The inspection of packaged foods and food intermediate during production by means of real-time acquired pictures is a well-known option when speaking of X-rays controls (Parisi 2013). On the other side, DIAP techniques do not concern the detection of foreign (and micro- or macroscopically evident) bodies in food products, while their application may be useful when speaking of food contamination (Correia and Mittal 1986; De Sitter and van de Haar 1998; Du and Sun 2004; Hui 2007; Leon et al. 2006; Maijala 2014; Mendoza et al. 2006; Parmee 1990; Upmann and Bonaparte 1999; Yam and Papadakis 2004).

DIAP advantages are dependent on the peculiar property or properties under investigation: in general, these features are chemical and physical, or microbiological variables with important information for the safety evaluation of foods, beverages and related containers. In detail, each chemical, physical or microbiological modification of foods and beverages (including the non-edible part of the packaged product, the packaging material) may potentially be analysed in terms of visually detectable modification of the food product without the need of destructive controls and contact-required techniques. This option has been verified with reference to foods (Parisi 2013) and food packaging materials (Gökmen and Süğüt 2007; Riva 2003). The analysis of digitally acquired pictures where colours have a distinctive role (as function of one or more specified qualitative or quantitative factors) can be made by means of reliable computed algorithms for online (in-process) and offline control (Parisi 2013).

The main advantage of DIAP systems is surely the possible evaluation of food products without destruction and contact, provided that a reliable relationship between the researched analyte and the obtained electronic 'answer' (the digitally acquired image) exists. In addition, simple instruments are required for these inspections: digital cameras, dedicated scanner instruments and personal computer software for image analysis and elaboration. On the other hand, the possible disadvantage of these techniques is always correlated with the inspection of superficial properties only. In other terms, each acquired image is only the reflection of emitted light in a reduced portion of the visible spectrum (380–780 nm) and it is referred explicitly to the external surface of products without penetration in the inner structure or layered network (where possible). Consequently, no inner controls are possible and reliable. Moreover, the basis of colorimetry applied in this inspection ambit is normally the comparison between a known sample and the unknown object; actually, the need of 'blank' samples could be seen as a distinctive advantage in some sectors and for different applications.

In brief, DIAP systems are based on the association of the perceived 'primary' colours by the human eye (blue, green and red). This association is defined as 'colour space' (Gökmen and Süğüt 2007; Riva 2003; Rosen et al. 2000).

The Reader is invited to consult the definition of colour spaces in more specific publications. However, it can be assumed here that the most studied and applied colour space—the L*a*b* model—appears at present as the most reliable choice, in accordance with the proposal of the Commission Internationale d'Eclairage in 1976. Three components define this colour space model (Parisi 2013):

(1) L* is named 'luminance' (values: 0–100). It means the amount of reflected visible light from the sample surface;
(2) a* is a dedicated parameter expressing the predominance of a specified chromatic tint (values: −120 to +120, where maximum and minimum numbers represent the totality of 'green' and 'red' colours, respectively) and
(3) b* is another dedicated parameter expressing the predominance of a specified chromatic tint (values: −120 to +120, where maximum and minimum numbers represent the totality of 'blue' and 'yellow' colours, respectively).

However, the definition of the red/green/blue (RGB) system is not sufficient. First of all, it should be mentioned that the mathematical definition of red, blue and green colours is made possible by means of peculiar lightness or luminance values (values: 0–255). In addition, the Munsell-based colour space system—also named Hue–Lightness–Saturation system—should be necessarily cited. However, simple comparative testing methods could also be carried out using only L*a*b* models when speaking of certain foods, beverages and related containers. The following sections are dedicated to the description of possible visual comparisons of food/beverage products and food packaging materials respectively in the L*a*b* colour space.

Before proceeding with practical descriptions, it should be remembered that digital acquisition can be reliable only provided that the following instruments are available (Riva 2003):

(1) Illumination source;
(2) Lamp angle (two illuminant sources may be recommended);
(3) Distance between camera and object;
(4) Digital device for acquisition and
(5) Processing software.

In addition, each acquired image is the association of a defined number of differently coloured pixels (bidimensional area sections). Each pixel (or radiant area in a peculiar place with space coordinates x and y) is mathematically identified by means of a determined light intensity (0–255). This value is the result of three different light intensity numbers related to red, green and blue radiations, respectively. As a result, each image portion may be analysed by processing software with the aim of determining:

(1) The global light intensity or 'grayscale' intensity (for a specified area or the whole image, between 0 and 255) and

(2) The light intensity for red, green and blue colours, respectively (please note that the normal red, green and blue radiations are emitted in the RGB system at 700, 540 and 460 nm, respectively).

Substantially, the comparison between 'blank' samples and unknown products can be performed by means of the comparison of related light intensity values (grayscale, red, green and blue colours). Consequently, the possible alteration of certain tints can be analysed and critically discussed, provided that a reliable relationship has been established between the above-mentioned chromatic modification and the cause (the variation of one chemical, physical or microbiological feature).

3.2   Image Analysis: Possible Applications for Thermally Treated Edible Foods and Containers

Foods and beverages are generally observed and evaluated on the market on the basis of a few appreciable parameters, including the chromatic appearance (Parisi 2013). Consequently, should a peculiar chemical or microbiological change determine an appreciable chromatic variation of the packaged food without concomitant interferences, the simple analysis of light intensity values at a specified wavelength would be useful.

Processed tomato sauces may offer a simple example for this type of examination. The main pigment for tomato products is known to be lycopene (deep-red tint); consequently, should this substance (typically found in fresh tomatoes in the all-*trans* configuration) be partially isomerised (from all-*trans* to *cis*-forms) or oxidised, the result should be very different in terms of tint, provided that the chromatic alteration depends only by lycopene loss (Shi and Le Maguer 2000).

Another option can be possible when speaking of peculiar products such as pasteurised *Harissa* sauces (Parisi et al. 2013). In this particular situation, the possible discoloration of products is not only ascribed to chemical modification of pigments into the product (index of drastic thermal treatment, inadequate raw material choice, use of acidified oils, etc.), but also to the transferal of a certain quantity of these pigments into the can coating enamel (on the inner side of the container). On the other side, this meshing phenomenon may modify irreversibly the inner white enamel of the container (Parisi 2012–2013). As a result, a digital acquisition of colorimetric modifications of the food or the packaging (or both objects) may be used with the aim of defining quality and quantitative modifications of selected products.

T1 sample – acquired picture

T2 sample – acquired picture

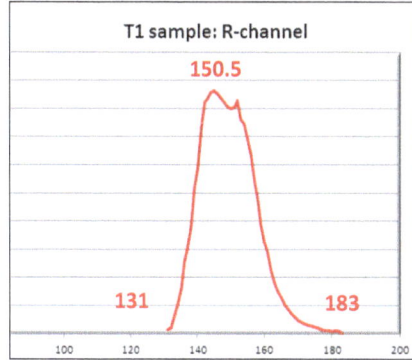

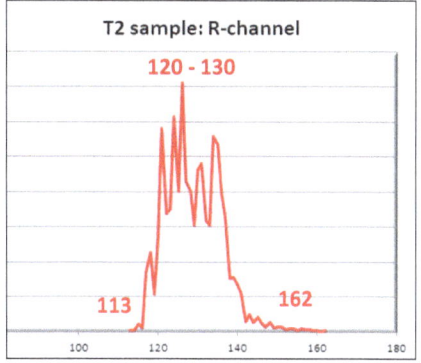

Fig. 3.1 A digital comparison between two sterilised canned tomato sauce products. The aim of the comparison is to evaluate the possible lycopene loss after sterilisation by means of the acquisition of two pictures related to different canned products, named T1 and T2 (tomato sauces). Two graphs are displayed under the related acquired images. Sample T1 appears to show a certain colour retention (range: 131–183) while T2 sample is more discoloured (range: 113–162)

A simulative example is provided here (Fig. 3.1): a comparison between two sterilised canned tomato sauce products. The aim of the comparison is to evaluate the possible lycopene loss after sterilisation by means of the acquisition of two pictures (tomato sauce).

Figure 3.1 shows a simulation of the situation where T1 and T2 are the two sampled canned foods, and two graphs are displayed under the related samples. Sample T1 appears to show a certain colour retention, provided that the reference for non-sterilised tomato sauce is represented by the following data:

(a) Minimum light intensity, red (R) channel (700 nm): 135;
(b) Maximum light intensity, red (R) channel (700 nm): 175 and
(c) Average light intensity, red (R) channel (700 nm): 153.

It has to be considered that X-axis represents the light intensity (0–255), while Y-axis represents the number of pixels showing a determined light intensity. As a result, each X-axis value is in relation with a determined pixel number on Y-axis, and this relationship allows evaluating the graph.

The comparison shows that T1 sample is close to the 'ideality' (or the 'blank' sample) because minimum, maximum and average light intensity values are 131, 138 and 150.5, respectively. On the other side, T2 sample 'is not good enough' if compared with T1 and the 'blank' reference because minimum, maximum and average values are 113, 162 and 120–130, respectively. In other terms, T1 sample is not discoloured if compared with blank results, while T2 is discoloured enough. The difference '153—125', where 125 is a rough medium value of average values for T2, can be used to determine a qualitative 'lycopene loss' index (LLI), on condition that this index measures the decrease of all pigments and this loss is calculated as lycopene diminution). The LLI index should be calculated in comparison with the ideal number (153) and expressed as percentage value. As a result, LLI for T1 would be higher than LLI for T1 (Eq. 3.1):

$$\text{LLI}_{\text{T1}} = \frac{(153 - 150.5)}{153} \times 100 = 1.6\% \tag{3.1}$$

On the other hand, LLI for T2 would be (Eq. 3.2):

$$\text{LLI}_{\text{T2}} = \frac{(153 - 125)}{153} \times 100 = 18.3\% \tag{3.2}$$

The above-mentioned qualitative comparison may be carried out without approximate indexes such as LLI. In fact, each graph obtained for the two samples can generally be obtained with the aim of calculating more specific information with reference to light intensity. The most common software for image processing and analysis can also calculate the number of pixels for each light intensity value. In other terms, should T1 sample be considered, the graph would display a 'reflection' curve (Parisi 2013; Parisi et al. 2013) where each light intensity value (range: 0–255) is associated with a specific pixel number. The analysis of the image obtained for sample T1 can give a list of values shown in Table 3.1.

As a result, the average light intensity (ALI) can be calculated in the R (red) channel as follows (Eq. 3.3):

$$\text{ALI} = \sum\nolimits_{i=1}^{255} (P_i \times L_i) \Big/ \sum\nolimits_{i=1}^{255} P_i, \tag{3.3}$$

where P_i is the number of pixels for the specified light intensity value L_i. The final result can show the average light intensity effectively emitted from the acquired image (once more, the range is 0–255). Consequently, T1 has an ALI value of 149, while T2 (Table 3.2) has an ALI of 129.

The comparison between T1 and T2 samples can be made by means of ALI values; in addition, the estimated LLI can be calculated in comparison with the 'ideal' ALI or the measured ALI for the blank reference. Should ALI for the blank

Table 3.1 The digital image analysis of the T1 sample

Light intensity (L_i)	Pixel number (P_i)	Light intensity (L_i)	Pixel number (P_i)	Light intensity (L_i)	Pixel number (P_i)
131	87	149	4040	167	438
132	113	150	4009	168	360
133	329	151	4029	169	298
134	560	152	4142	170	245
135	846	153	3815	171	214
136	1263	154	3736	172	175
137	1552	155	3472	173	160
138	1916	156	3182	174	141
139	2603	157	2729	175	122
140	2942	158	2381	176	113
141	3640	159	1960	177	83
142	4112	160	1655	178	59
143	4192	161	1464	179	68
144	4290	162	1124	180	56
145	4322	163	904	181	63
146	4281	164	749	182	45
147	4192	165	642	183	24
148	4120	166	510		

The most common software for image processing and analysis can calculate the number of pixels (P_i) for each light intensity value (L_i). Each 'light intensity' value is associated with a specific pixel number. As a result, the graph obtained by sample T1 (Fig. 3.1) can give a list of values. The average light intensity (ALI) is calculated in the R (red) channel. T1 has an ALI value of 149

sample be 153, the estimated lycopene loss should be calculated as shown by Eq. 3.1, and the same thing should be done for T2. Obtained results would be:

(a) LLI_{T1} = 2.6% and
(b) LLI_{T2} = 15.7%.

The above-mentioned experience can be also performed for can coatings after sterilisation with tomato sauces. After sterilisation, defective coatings may be irreversibly penetrated by organic red pigments. The acquired digital image of white enamel-coated inner side of metal cans can be analysed and compared for comparison purposes (experimental studies—routine evaluation should be not recommended), provided that colorimetric modifications concern the incorporation of 'red' pigments (Parisi et al. 2013).

Table 3.2 The digital image analysis of the T2 sample

Light intensity (L_i)	Pixel number (P_i)	Light intensity (L_i)	Pixel number (P_i)	Light intensity (L_i)	Pixel number (P_i)
131	87	149	4040	167	438
132	113	150	4009	168	360
133	329	151	4029	169	298
134	560	152	4142	170	245
135	846	153	3815	171	214
136	1263	154	3736	172	175
137	1552	155	3472	173	160
138	1916	156	3182	174	141
139	2603	157	2729	175	122
140	2942	158	2381	176	113
141	3640	159	1960	177	83
142	4112	160	1655	178	59
143	4192	161	1464	179	68
144	4290	162	1124	180	56
145	4322	163	904	181	63
146	4281	164	749	182	45
147	4192	165	642	183	24
148	4120	166	510		

The most common software for image processing and analysis calculates the number of pixels (P_i) for each light intensity value (L_i). Each 'light intensity' value is associated with a specific pixel number. As a result, the graph obtained by sample T2 (Fig. 3.1) can give a list of values. The average light intensity (ALI) is calculated in the R (red) channel. T2 has an ALI value of 129

References

AACC International (2009) Thermophilic spore counts (Total Aerobic, Flat-Sour, H₂S, Non-H₂S Anaerobic). Approved methods of analysis, 11th edn. Method 42-40-01. American Association of Cereal Chemists (AACC) International, St. Paul. https://doi.org/10.1094/aaccintmethod-42-40.01

Andreis G, Ottaviani F (eds) (2002) Manuale della Sicurezza Microbiologica degli Alimenti e delle Acque. Oxoid Spa, Garbagnate Milanese

Bhagat A, Caruso G, Micali M, Parisi S (2016) Foods of non-animal origin. SpringerBriefs in Chemistry of Foods, Springer International Publishing, Cham. https://doi.org/10.1007/978-3-319-25649-8

Blaschek HP (1999) Clostridium. *Clostridium Perfringens*. Encycl Food Microbiol 433–438. https://doi.org/10.1006/rwfm.1999.0375

Bordonaro D (2012) Il Controllo Ufficiale dei M.O.C.A. e le Linee Guida della Regione Piemonte. In: Proceedings of the Corso in Alta Formazione in Legislazione Alimentare. University of Piemonte Orientale, Alessandria

Chen Z (2017) Microbial inactivation in foods by ultrasound. J Food Microbiol Saf Hyg 02(1):1–2. https://doi.org/10.4172/2476-2059.1000e102

Chen Y, Yu LJ, Rupasinghe HV (2013) Effect of thermal and non-thermal pasteurisation on the microbial inactivation and phenolic degradation in fruit juice: a mini-review. J Sci Food Agric 93(5):981–986. https://doi.org/10.1002/jsfa.5989

Correia LR, Mittal GS (1986) Application of computer-aided digital image analysis in food processing. Can Inst Food Sci Technol J 19(4): xlvi. https://doi.org/10.1016/s0315-5463(86)71600-3

de Fátima Pocas M, Hogg T (2007) Exposure assessment of chemicals from packaging materials in foods: a review. Trend Food Sci Technol 18(4):219–230. https://doi.org/10.1016/j.tifs.2006.12.008

Delia S, Laganà P, Parisi S (2005) Materiali e metodi di confezionamento nella conservazione dei prodotti alimentari refrigerati. In: Proceedings of the XIV Conferenza Nazionale "Microbiologia degli alimenti conservati in stato di refrigerazione". University of Bologna, Bologna

De Sitter H, van de Haar S (1998) Governmental food inspection and HACCP. Food Control 9(2–3):131–135. https://doi.org/10.1016/s0956-7135(98)00083-8

Dotzauer C, Ehrmann MA, Vogel RF (2002) Occurrence and detection of *Thermoanaerobacterium* and *Thermoanaerobacter* in canned food. Food Technol Biotechnol 40(1):21–26

Downes FP, Ito K (eds) (2001) Compendium of methods for the microbiological examination of foods, 4th edn. American Public Health Association, Washington

Du CJ, Sun DW (2004) Recent developments in the applications of image processing techniques for food quality evaluation. Trends Food Sci Technol 15(5):230–249. https://doi.org/10.1016/j.tifs.2003.10.006

Evancho GM, Walls I (2001) Aciduric Flat Sour Sporeformers. Compendium of methods for the microbiological examination of foods. American Public Health Association, Washington, D.C, pp 239–244. https://doi.org/10.2105/9780875531755ch24

Fields ML (1970) The flat sour bacteria. Adv Food Res 18:163–217. https://doi.org/10.1016/s0065-2628(08)60370-5

Gökmen V, Süğüt I (2007) Expert commentary—computer-vision based analysis of color as a tool for food process control. In: Pletney VN (ed) Focus on food engineering research and developments. Nova Science Publishers Inc, New York

Gooch JW (2011) Flat sour spoilage. In: Good JW (ed) Encyclopedic dictionary of polymers, pp 893–893. Springer, New York. https://doi.org/10.1007/978-1-4419-6247-8_13764

Gupta R, Balasubramaniam VM (2012) High-pressure processing of fluid foods. In: Cullen PJ, Tiwari BK, Valdramidis VP (eds) Novel thermal and non-thermal technologies for fluid foods. Academic Press, London, Waltham, and San Diego, pp 109–133. https://doi.org/10.1016/b978-0-12-381470-8.00005-0

Hui YH (2007) Food establishment inspection. In: Hui YH, Chandan RC, Clark S, Cross NA, Dobbs JC, Hurst WJ, Nollet LML, Shimoni E, Sinha N, Smith EB, Surapat S, Toldrá F, Titchenal A (eds) Handbook of food products manufacturing. Wiley, Hoboken. https://doi.org/10.1002/9780470113554.ch13

Icier F (2012) Ohmic heating of fluid foods. In: Cullen PJ, Tiwari BK, Valdramidis VP (eds) Novel thermal and non-thermal technologies for fluid foods. Academic Press, London, Waltham, and San Diego, pp 305–367. https://doi.org/10.1016/b978-0-12-381470-8.00011-6

Ikem A, Egiebor NO (2005) Assessment of trace elements in canned fishes (mackerel, tuna, salmon, sardines and herrings) marketed in Georgia and Alabama (United States of America). J Food Compos Anal 18(8):771–787. https://doi.org/10.1016/j.jfca.2004.11.002

ISO (2013) Prerequisite programmes on food safety—Part 4: food packaging manufacturing. International Organization for Standardization (ISO), Geneva

Jay JM, Loessner MJ, Golden DA (eds) (2008a) Modern food microbiology, 7th edn. Springer Science & Business Media, New York, pp 301–350

Jay JM, Loessner MJ, Golden DA (eds) (2008b) Modern food microbiology, 7th edn. Springer Science & Business Media, New York, pp 709–726

Johnson EA (1999) *Clostridium—Clostridium Botulinum*. Encyclopedia Food Microbiol 458–463. https://doi.org/10.1006/rwfm.1999.0395

Jongyingcharoen JS, Ahmad I (2014) Thermal and non-thermal processing of functional foods. In: Noomhorm A, Ahmad I, Anal AK (eds) Functional foods and dietary supplements: processing effects and health benefits. Wiley, Chichester. https://doi.org/10.1002/9781118227800.ch11

Leon K, Mery D, Pedreschi F, Leon J (2006) Color measurement in L∗a∗b∗ units from RGB digital images. Food Res Int 39(10):1084–1091. https://doi.org/10.1016/j.foodres.2006.03.006

Maijala R (2014) Scientific risk assessment-basis for food legislation. In: Ninios T, Lundén J, Korkeala H, Fredriksson-Ahomaa M (eds) Meat inspection and control in the slaughterhouse. Wiley, Chichester. https://doi.org/10.1002/9781118525821.ch26

Mañas P, Pagán R (2005) Microbial inactivation by new technologies of food preservation. J Appl Microbiol 98(6):1387–1399. https://doi.org/10.1111/j.1365-2672.2005.02561.x

Matsuda N, Komaki M, Ichikawa R, Gotoh S (1985) Aerobic and facultative anaerobic spore-forming bacteria isolated from spoiled canned foods. J Jpn Soc Food Sci Technol 32(6):399–406. https://doi.org/10.3136/nskkk1962.32.6_399

McClane BA (2007) Clostridium perfringens. In: Doyle M. Beuchat L (eds) Food microbiology: fundamentals and frontiers, 3rd edn. ASM Press, Washington, D.C, pp 423–444. https://doi.org/10.1128/9781555815912.ch19

Mendoza F, Dejmek P, Aguilera JM (2006) Calibrated color measurements of agricultural foods using image analysis. Postharvest Biol Technol 41(3):285–295. https://doi.org/10.1016/j.postharvbio.2006.04.004

Miyamoto K, Nagahama M (2016) Clostridium: food poisoning by Clostridium perfringens. In: Caballero B, Finglas PM, Toldrá F (eds) Encyclopedia of food and health. Academic Press, Oxford, pp 149–154. https://doi.org/10.1016/b978-0-12-384947-2.00171-9

Notermans SHW (1999) Clostridium—detection of neurotoxins of Clostridium Botulinum. Encyclopedia Food Microbiol 463–466. https://doi.org/10.1006/rwfm.1999.0400

Oliveira SR, Cruz RM, Vieira MC, Silva CL, Gaspar MN (2009) Enterococcus faecalis and Pseudomonas aeruginosa behaviour in frozen watercress (Nasturtium officinale) submitted to temperature abuses. Int J Refrig 32(3):472–477. https://doi.org/10.1016/j.ijrefrig.2008.08.002

Olson KE, Sorrells KM (2015) 26. Thermophilic flat sour sporeformers. In: Salfinger Y, Tortorello ML (eds) Compendium of methods for the microbiological examination of foods. American Publih Health Association, Wagshington, D.C. https://doi.org/10.2105/mbef.0222.031

Ottaviani F (ed) (1996) Microbiologia dei Prodotti di Origine Vegetale—Ecologia ed Analisi Microbiologica. Chiriotti Editori, Pinerolo

Parisi S (2004) Alterazioni in imballaggi metallici termicamente processati. Gulotta Press, Palermo

Parisi S (2009) Intelligent packaging for the food industry. In: Gardiner F, Carter E (eds) Polymer electronics—a flexible technology. Smithers Rapra Technology Ltd, Shawbury

Parisi S (2012) Food packaging and food alterations: the user-oriented approach. Smithers Rapra Technology, Shawbury

Parisi S (2013) Food industry and packaging materials performance-oriented guidelines for users. Smithers Rapra Technology, Shawbury

Parisi S (2016a) The world of foods and beverages today: globalization, Crisis management and future perspectives. The Economist Group, London and New York. Available http://learning.ly/products/the-world-of-foods-and-beverages-today-globalization-crisis-management-and-future-perspectives. Accessed 29 June 2017

Parisi S (2016b) Alberto Escarpa, María Cristina González and Miguel Angel López (eds) Agricultural and food electroanalysis. Anal Bioanal Chem 408:2185–2186. https://doi.org/10.1007/s00216-016-9347-9

Parisi S, Laganà P, Gioffrè ME, Minutoli E, Delia S (2013) Problematiche emergenti di sicurezza alimentare. Prodotti etnici e autenticità. In: Proceedings of the XXIV Congresso Interreggionale Siculo-Calabro SItI, Palermo

Parmee R (1990) X-rays give in-depth inspection. Sens Rev 10(2):84–86. https://doi.org/10.1108/eb007818

Pflug IJ, Davidson PM, Holcomb RG (1981) Incidence of canned food spoilage at the retail level. J Food Prot 44(9):682–685. https://doi.org/10.4315/0362-028X-44.9.682

Piergiovanni L, Limbo S (2016) Food packaging materials. SpringerBriefs in Chemistry of Foods, Springer International Pub-lishing, Cham. https://doi.org/10.1007/978-3-319-24732-8

Pisanello D (2014) Chemistry of foods: EU legal and regulatory approaches. SpringerBriefs in Chemistry of Foods, Springer International Publishing, Heidelberg. https://doi.org/10.1007/978-3-319-03434-8

Rastogi NK (2012) Infrared heating of fluid foods. In: Cullen PJ, Tiwari BK, Valdramidis VP (eds) Novel thermal and non-thermal technologies for fluid foods. Academic Press, London, Waltham, and San Diego, pp 411–432. https://doi.org/10.1016/b978-0-12-381470-8.00013-x

Richardson KC (1972) Microbial spoilage in Australian canned foods, 1955–68. Food Technol Aust 24:106–107

Riva M (2003) Il colore degli alimenti e la sua misurazione. Dipartimento Di Scienze E Tecnologie Alimentari E. Microbiologiche, University of Milan, Milan

Rosen MR, Fairchild MD, Johnson GM, Wible DR (2000) Color Management within a spectral image visualization tool. In: Proceedings of the IS&T/SID eighth color imaging conference: color science and engineering, systems, technologies and applications, Springfield

Sahin S, Sumnu SG (2006) Thermal properties of foods. In: Sahin S, Sumnu SG (eds) Physical properties of foods, pp 107–155. https://doi.org/10.1007/0-387-30808-3_3

Savov AV, Kouzmanov GB (2009) Food quality and safety standards at a glance. Biotechnol Biotechnol Equip 23(4):1462–1468. https://doi.org/10.2478/V10133-009-0012-8

Shi J, Le Maguer M (2000) Lycopene in tomatoes: chemical and physical properties affected by food processing. Crit Rev Food Sci Nutr 40(1):1–42. https://doi.org/10.1080/10408690091189275

Tiwari BK, Mason TJ (2012) In: Cullen PJ, Tiwari BK, Valdramidis VP (eds) Novel Thermal and non-thermal technologies for fluid foods. Academic Press, London, Waltham, and San Diego, pp 135–165. https://doi.org/10.1016/b978-0-12-381470-8.00006-2

Upmann M, Bonaparte C (1999) Rapid methods for food hygiene inspection. Encyclopedia Food Microbiol 1887–1895. https://doi.org/10.1006/rwfm.1999.1320

Vasseur C, Baverel L, Hebraud M, Labadie J (1999) Effect of osmotic, alkaline, acid or thermal stresses on the growth and inhibition of *Listeria monocytogenes*. J Appl Microbiol 86(3):469–476. https://doi.org/10.1046/j.1365-2672.1999.00686.x

Yam KL, Papadakis SE (2004) A simple digital imaging method for measuring and analyzing color of food surfaces. J Food Eng 61(1):137–142. https://doi.org/10.1016/s0260-8774(03)00195-x

Chapter 4
Canned Tomato Sauces and Beans: Industrial Processes

Angela Montanari, Caterina Barone, Michele Barone
and Anna Santangelo

Abstract The modern industry of canned foods is correlated with a selected portion of commercial products: fruits and vegetable foods, meat and meat products, shellfish and fish. Two of these products, tomato sauces and baked beans, may be described in detail with the aim of highlighting the importance of several processing steps: the packaging (and the role of metal cans) with quality control procedures, and general defects of canned tomato sauces and baked beans, with a general description of traceability needs. Good or excellent microbiological quality has to be assured when speaking of vegetables, water for washing operations and production equipment. Basic controls on metal cans should consider and evaluate the minimum positive features expected for similar rigid containers. Finally, the hygienic production of canned foods has to be performed by means of adequate good manufacturing practices. In this ambit, the importance of traceability has to be considered when speaking of foods and packaging materials.

Keywords Canned bean · Coating · Raw material · Solid waste
Technological suitability · Tomato sauce · Traceability

4.1 Canned Tomato Sauces and Beans in the Modern Food Industry

The modern industry of canned foods is correlated with a selected portion of commercial products: fruits and vegetable foods, meat and meat products, shellfish and fish (Larousse and Brown 1997; Simpson 2012). Anyway, the above-mentioned products are always 'processed foods' because thermal treatments for the realisation of the final packaged product imply a certain difference between the original food and the final product (Parisi 2004; Simpson 2012). In addition, canned foods have a long history (two centuries of industrial improvements of the initial preserved products) and excellent features such as extended shelf-life periods, a certain cheapness if compared with similar (non-canned) food products and low production costs (Brown 2000; Carslaw and Jaeger 1959; Durance 1997;

© The Author(s), under exclusive licence to Springer International Publishing AG,
part of Springer Nature 2018
A. Montanari et al., *Thermal Treatments of Canned Foods*,
Chemistry of Foods, https://doi.org/10.1007/978-3-319-74132-1_4

Engehnan and Sani 1983; Hersom and Hulland 1981; Jen et al. 1971; Lenz and Lund 1977; Lund 1982; Mishkin et al. 1982; Pflung and Odlaugh 1978; Simpson 2012; Teixeira 1971).

Two of these products, tomato sauces and baked beans, may be described in detail with the aim of highlighting the importance of several processing steps:

(a) The packaging (and the role of metal cans) with quality control procedures (Sect. 4.2), and
(b) General defects of canned tomato sauces and baked beans, with a general description of traceability needs (Sect. 4.3).

Raw materials (vegetables) have to be prepared adequately and the peculiar system depends on the initial raw material and the preservation/packaging method (Simpson 2012). The Reader is invited to read works that are more exhaustive on processing steps for the industry of canned foods and fruits/vegetables in particular (Gómez-Sánchez 2007; Jay et al. 2008a, b; NFPA/CMI Container Integrity Task Force 1984; Schmitt 1966). Two examples can be briefly shown here with concern to the same can typology: tomato sauces (28–30 °Brix) and canned beans.

With relation to tomato sauces, harvested raw materials of good quality are immediately subjected to the following steps as shown in Fig. 4.1 (Barringer 2004; Larousse and Brown 1997; Luh and Kean 1988; Simpson 2012):

(1)   Reception,
(2)   Storage and inspection,
(3)   Sorting/grading,
(4)   Washing,
(5)   Automatic/mechanical peeling,
(6)   Cutting and trimming,
(7)   Hot-break (85–100 °C) step for the inactivation of polygalacturonase and pectin methylesterase enzymes,
(8)   Pulping and refining (also named juice extraction),
(9)   Deaeration,
(10) Homogenization,
(11) Evaporation/concentration (48–82 °C),
(12) Volumetric filling,
(13) Seaming,
(14) Thermal treatment (pasteurisation, sterilisation, high-pressure processing, ohmic or microwave heating, etc.),
(15) Cooling,
(16) Labelling and
(17) Final storage.

On the other side, the industrial production of canned beans is performed as shown in Fig. 4.2 (Larousse and Brown 1997; Schoeninger et al. 2017; Tiwari and Singh 2012; White and Howard 2012):

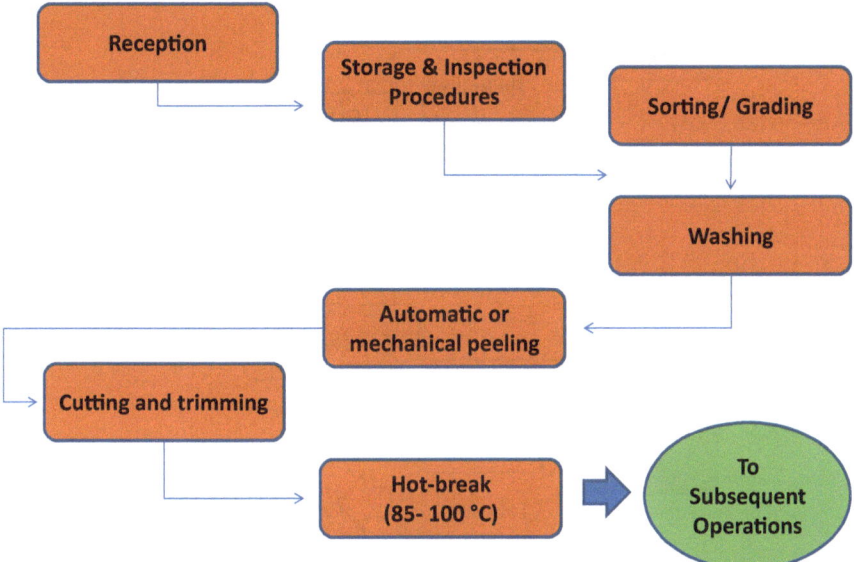

Fig. 4.1 Raw vegetables have to be prepared adequately in the food industry. The initial operations are quite similar in many environments. It has to be noted that washing is not generally required when speaking of legumes; the same thing is observed for peeling and cutting/trimming processes (this operation is not always carried out when speaking of legumes). After blanching, processing steps depend on the peculiar type of product (tomato sauce, canned beans, etc.)

(a) Cleaning,
(b) Grain examination,
(c) Hydration (different systems may be used depending on industrial preferences and the type of processed bean),
(d) Grain classification,
(e) Blanching (this step is not always performed after hydration),
(f) Gravimetrical or volumetric filling in aluminium or tin cans (please note cans have to be preliminarily washed),
(g) Addition of liquid media (brine made of sodium chloride sugar, and/or calcium chloride and ethylenediaminetetraacetic acid, or tomato sauce for baked beans),
(h) Can seaming,
(i) Thermal processing,
(j) Cooling,
(k) Labelling and
(l) Final storage.

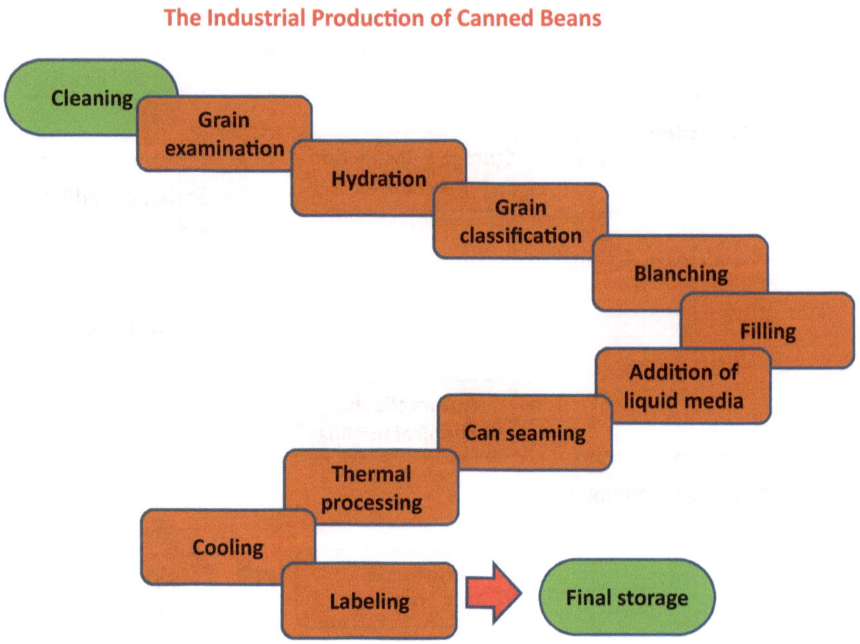

The Industrial Production of Canned Beans

Cleaning

Grain examination

Hydration

Grain classification

Blanching

Filling

Addition of liquid media

Can seaming

Thermal processing

Cooling

Labeling

Final storage

Fig. 4.2 The production of canned beans in the modern industry

4.2   Industrial Processes: Metal Cans and Quality Control Protocols Concerning the Packaging in Connection with Food

The most used container type for canned tomato products and beans is historically the tinplate-made 'metal can' (Palmieri et al. 2004; Parisi 2012). Interestingly, aluminium alloys can be interesting enough for several sectors such as canned beans (Schoeninger et al. 2017); however, the traditional 'tin can' with or without beadings is still well known. Other containers—glass jars, plastic packaging materials, etc.—can be used, but this chapter is dedicated to canned foods in tinplate-made cans.

The metallic surface has to be covered with organic coatings (lacquers) on both sides of the metal can and ends, with optional white enamels on the inner side. In general, the so-called 'three-piece can' (composed of two ends and a cylindrical body) is used although the 'two-piece can' (one aluminium-made body and one end only) can be interesting because of the absence of welded areas on the cylindrical body (a dangerous section without adequate coating protection). Ends have to be applied (Barbieri and Rosso 1990) by the initial can manufacturer (three-piece cans only), and by the final food producer (two-piece cans, one end only, and three-piece cans, the last end).

Quality control procedures in the canning industry are essentially dedicated to the performance of the entire food product (food/packaging system). The analytical approach should be considered when speaking of visually appreciable defects: in this ambit, technologies such as X-rays inspection and digital image and processing evaluations (Chap. 3) may be potentially useful (Correia and Mittal 1986; De Sitter and van de Haar 1998; Du and Sun 2004; Hui 2007; Leon et al. 2006; Maijala 2014; Mendoza et al. 2006; Parmee 1990; Upmann and Bonaparte 1999; Yam and Papadakis 2004). On the one side, basic controls on metal cans should consider and evaluate the minimum positive features expected for similar rigid containers (Larousse and Brown 1997; Parisi 2013), as shown in Fig. 4.3:

(a) Good thickness and hardness of the body can. Control: superficial hardness of steel and thickness;
(b) Absence of metallic defects. Examinations can be made by means of optical systems;
(c) Excellent resistance offered by body beadings;
(d) Good axial load resistance of the whole can;
(e) Good double-seam parameters (closure of ends) and
(f) Excellent covering of metallic surfaces by means of organic coatings and/or enamels, where requested (absence of dewettings, scratches, etc.). Testing methods are simple chemical testing methods with the aim of detecting uncovered tin. An example is the reaction of copper sulphate solution on uncoated tin-covered surfaces, with copper deposition (red pinpoints).

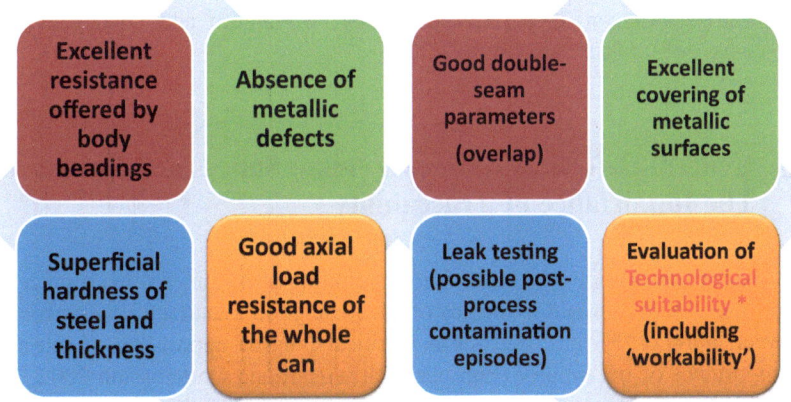

Fig. 4.3 General quality control procedures in the modern canning industry

These quality controls are specific for the industry of canmakers; however, food industry should test and evaluate in strong cooperation with the canmaker the so-called 'technological suitability' (Sect. 3.1), including workability, of the initial metal cans before using it for the intended use (Bordonaro 2012). Some example is offered in this book (Chap. 3). The interested Reader is invited to read more specific texts with concern to quality control procedures for metal can analysis and evaluation. A dedicated control plan should be performed on a statistical basis because of the huge amount of single cans into a specified lot, depending on the dimension of a single container (Barbieri and Rosso 1990). The seam inspection at regular intervals is a critical step when speaking of proper closure in metal cans. Double seam has to be evaluated with reference to correct thickness, height (width) and countersink, with the aim of evaluating overlap performances (Larousse and Brown 1997). Finally, the 'leak testing' may be required on cans because of possible post-process contamination episodes (Jay et al. 2008c).

With relation to other quality control testing methods, the importance of raw materials has been highlighted. Good or excellent microbiological quality has to be assured when speaking of vegetables, water for washing operations and production equipment; otherwise, thermal treatment strategies could fail because of the direct relationship between initial (known) data concerning food contamination and operative conditions (time, temperature). The concept of commercial sterility should be considered with attention (Augusto et al. 2014; Codex Alimentarius Commission 1979; Costa 2003; da Silva et al. 2013; Hall 1971; Heinz and Hautzinger 2007; Jay et al. 2008a; Membré and van Zuijlen 2011; Nightingale and Stallings 1986; Schmitt 1966). 'Commercially sterile' canned foods are expected to remain stable for extended storage periods. On the other side, the term 'absolute sterility' does not seem to have practical meanings at present. For these reasons, defects such as putrefaction and flat-sour effects should be always considered (AACC International 2009; Blaschek 1999; Evancho and Walls 2001; Fields 1970; Gooch 2011; Johnson 1999; McClane 2007; Miyamoto and Nagahama 2016; Notermans 1999; Olson and Sorrells 2015).

4.3   General Defects of Canned Tomato Sauces and Beans: The Importance of Traceability

The hygienic production of canned foods, including also canned tomato sauces and beans, has to be performed by means of adequate good manufacturing practices, in the broad ambit of 'Hazard Analysis and Critical Control' approaches. In this ambit, the importance of traceability (European Parliament and Commission 2002, 2004; Parisi 2012) has to be considered when speaking of foods and packaging materials. Anyway, the basis of a good traceability system (Fig. 4.4) should consider (Mania et al. 2016a, b)

The traceability in practice

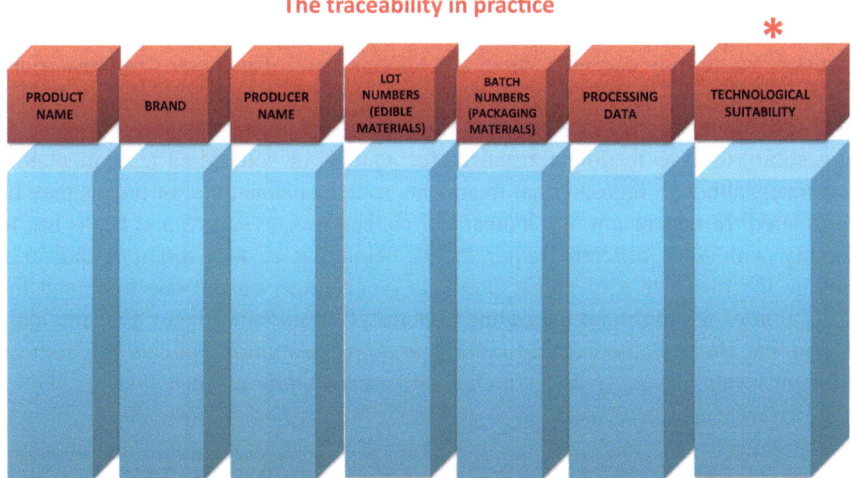

* **Cooperation between the Food Industry and the Canmaker**

Fig. 4.4 The main pillars of traceability systems in the modern Industry. Important requirements such as the 'technological suitability'—the compliance of food packaging materials in relation to specific norms and mandatory regulations and the intended use(s)—should be defined and correctly evaluated in strong cooperation between the food producer and the canmaker

(a) Name of the product;
(b) Specific brand;
(c) Name of Producer (food manufacturer and food packaging producer);
(d) Identification of lot numbers for raw materials, intermediate foods and final foods (it should be also good to subdivide and label products in sub-lot sections because of the supposed variability in production steps);
(e) Identification of batch numbers for packaging materials;
(f) Record of processing data (example: thermal values, times for different cycles, etc.) and
(g) Preventive evaluation of the so-called 'technological suitability' and workability.

This approach is extremely useful at present because of new and emerging issues such as the control of allergens in foods and correct labelling, in accordance with the Food Safety Modernization Act and similar laws in the United States of America (Parisi 2016). Apart these matters, currently observed defects on certain canned tomato sauces and beans are often correlated with the detection of pigment alterations, suspect browning effects, abnormal food oxidation and pH alterations (Schoeninger et al. 2017). On the other side, the commercial evaluation for canned beans concerns mainly headspace determination, evaluation of gross, net and drained weights, vacuum measurement, agglutination and colorimetric performances of bean grains and liquid media (Hosfield et al. 1984; Schoeninger et al. 2017; White and Howard 2012).

Another concern is correlated with tomato solid waste, although the same problem may be observed with other vegetables. With reference to the tomato industry, waste may exceed 15% of the total amount of processed raw materials. Consequently, the reuse of similar byproducts could be useful (Fernández-Gómez et al. 2010). In general, waste solids are treated by means of anaerobic systems (digestion), used as feeding substances for animals or considered as fermentation substrates, although high-thermal treatments such as incineration of biogas may be considered. In conclusion, the industry of canned tomato sauces and beans has to comply with very different requirements belonging to two different industrial ambits: the production of edible products (vegetables) on the one side, and the realisation of reliable food packaging containers on the other hand. For this main reason, the study of the canning industry requires a multidisciplinary research on different levels, including also commercial and economic aspects.

References

AACC International (2009) Thermophilic spore counts (Total Aerobic, Flat-Sour, H_2S, Non-H_2S Anaerobic). Approved methods of analysis, 11th edn. Method 42-40-01. American Association of Cereal Chemists (AACC) International, St. Paul. https://doi.org/10.1094/aaccintmethod-42-40.01

Augusto PED, Tribst AAL, Cristianini M (2014) Thermal processes—commercial sterility (Retort). In: Batt CA (ed) Encyclopedia of food microbiology. Academic Press, Cambridge, pp 567–576. https://doi.org/10.1016/b978-0-12-384730-0.00405-5

Barbieri G, Rosso S (1990) Il controllo dei contenitori di banda stagnata per conserve alimentari. Stazione Sperimentale per l'Industria delle Conserve Alimentari in Parma, Parma parisi

Barringer SA (2004) Vegetables: tomato processing. In: Scott Smith J, Hui YH (eds) Food processing: principles and applications. Blackwell Publishing, Ames and Oxford

Blaschek HP (1999) Clostridium. *Clostridium perfringens*. In: Encyclopedia of food microbiology, pp 433–438. https://doi.org/10.1006/rwfm.1999.0375

Bordonaro D (2012) Il Controllo Ufficiale dei M.O.C.A. e le Linee Guida della Regione Piemonte. In: Proceedings of the Corso in Alta Formazione in Legislazione Alimentare, University of Piemonte Orientale, Alessandria, 4 May 2012

Brown KL (2000) Control of bacterial spores. Brit Med Bull 56(1):158–171. https://doi.org/10.1258/0007142001902860

Carslaw HS, Jaeger JC (1959) Conduction of heat in solids. Clarendon Press, Oxford

Codex Alimentarius Commission (1979) Code of hygienic practice for low and acidified low acid canned foods, CAC/RCP 23-1979. The Codex Alimentarius Commission, The Food and Agriculture Organization of the United Nations, Rome

Correia LR, Mittal GS (1986) Application of computer-aided digital image analysis in food processing. Can Inst Food Sci Technol J 19(4):xlvi. https://doi.org/10.1016/s0315-5463(86)71600-3

Costa R (2003) Commercial food service establishments: the principles of modern food hygiene. In: Schmidt RH, Rodrick GE (eds) Food safety handbook, pp 453–522. Wiley, Hoboken, NJ, USA. https://doi.org/10.1002/047172159X.ch26

Da Silva N, Taniwaki M, Junqueira V, De Arruda Silveira N, Da Silva Do Nascimento M, Gomes R (2013) Microbiological examination methods of food and water: a laboratory manual. CRC Press/Balkema, Taylor & Francis Group, pp 311–333. https://doi.org/10.1201/b13740-24

De Sitter H, van de Haar S (1998) Governmental food inspection and HACCP. Food Control 9(2–3):131–135. https://doi.org/10.1016/s0956-7135(98)00083-8

Du CJ, Sun DW (2004) Recent developments in the applications of image processing techniques for food quality evaluation. Trends Food Sci Technol 15(5):230–249. https://doi.org/10.1016/j.tifs.2003.10.006

Durance TD (1997) Improving canned food quality with variable retort temperature process. Trends Food Sci Technol 8(4):113–118. https://doi.org/10.1016/S0924-2244(97)01010-8

Engehnan MS, Sani RL (1983) Finite-element simulation of an inpackage pasteurization process. Numer Heat Transf 6(1):41–54. https://doi.org/10.1080/01495728308963073

European Parliament and Council (2002) Regulation (EC) No 178/2002 of the European Parliament and of the Council of 28 January 2002 laying down the general principles and requirements of food law, establishing the European Food Safety Authority and laying down procedures in matters of food safety. Off J Eur Comm L31:1–24

European Parliament and Council (2004) Regulation (EC) No 1935/2004 of the European Parliament and of the Council of 27 October 2004 on materials and articles intended to come into contact with food and repealing Directives 80/590/EEC and 89/109/EEC. Off J L338:4–244

Evancho GM, Walls I (2001) Aciduric flat sour sporeformers. In: Compendium of methods for the microbiological examination of foods. American Public Health Association, Washington, DC, pp 239–244. https://doi.org/10.2105/9780875531755ch24

Fernández-Gómez MJ, Nogales R, Insam H, Romero E, Goberna M (2010) Continuous-feeding vermicomposting as a recycling management method to revalue tomato-fruit wastes from greenhouse crops. Waste Manag 30(12):2461–2468. https://doi.org/10.1016/j.wasman.2010.07.005

Fields ML (1970) The flat sour bacteria. Adv Food Res 18:163–217. https://doi.org/10.1016/s0065-2628(08)60370-5

Gómez-Sánchez A (2007) Microorganismos de importancia en el tratamiento térmico de alimentos ácidos y de alta acidez. Temas Selectos Ing Aliment 1:24–32

Gooch JW (2011) Flat sour spoilage. In: Good JW (ed) Encyclopedic dictionary of polymers. Springer, New York, pp 893–893. https://doi.org/10.1007/978-1-4419-6247-8_13764

Hall RC (1971) Simple test to predict commercial sterility of heated food products. J Milk Food Technol 34(4):196–197. https://doi.org/10.4315/0022-2747-34.4.196

Heinz G, Hautzinger P (2007) Meat processing technology for small to medium scale producers. Food and Agriculture Organization of the United Nations, Regional Office for Asia and the Pacific, Bangkok. ISBN: 978-974-7946-99-4. Available http://www.fao.org/3/a-ai407e.pdf. Accessed 17 Nov 2017

Hersom AC, Hulland ED (1981) Canned foods: thermal processing and microbiology, 7th edn. Chemical Publishing Company, New York

Hosfield CL, Ghaderi A, Uebersax MA (1984) A factor analysis of yield and sensory and physicochemical data from tests used to measure culinary quality in dry edible beans. Can J Plant Sci 64(2):285–293. https://doi.org/10.4141/cjps84-042

Hui YH (2007) Food establishment inspection. In: Hui YH, Chandan RC, Clark S, Cross NA, Dobbs JC, Hurst WJ, Nollet LML, Shimoni E, Sinha N, Smith EB, Surapat S, Toldrá F, Titchenal A (eds) Handbook of food products manufacturing. Wiley, Hoboken. https://doi.org/10.1002/9780470113554.ch13

Jay JM, Loessner MJ, Golden DA (eds) (2008a) Modern food microbiology, 7th edn. Springer Science & Business Media, New York, pp 415–441

Jay JM, Loessner MJ, Golden DA (eds) (2008b) Modern food microbiology, 7th edn. Springer Science & Business Media, New York, pp 567–590

Jay JM, Loessner MJ, Golden DA (eds) (2008c) Modern food microbiology, 7th edn. Springer Science & Business Media, New York, pp 457–472

Jen Y, Manson JE, Stumbo CR, Zahradnik JW (1971) A procedure for estimating sterilization of and quality factor degradation in thermally processed foods. J Food Sci 36(4):693–698. https://doi.org/10.1111/j.1365-2621.1971.tb15164.x

Johnson EA (1999) Clostridium—*Clostridium botulinum*. In: Encyclopedia of food microbiology, pp 458–463. https://doi.org/10.1006/rwfm.1999.0395

Larousse J, Brown BE (eds) (1997) Food canning technology. Wiley-VCH Inc., New York, Chichester, Weinheim, Brisbane, Singapore, Toronto

Lenz MK, Lund DB (1977) The lethality-Fourier number method: experimental verification of a model for calculating average quality factor retention in conduction-heated, canned foods. J Food Sci 42(4):989–996. https://doi.org/10.1111/j.1365-2621.1977.tb12652.x

Leon K, Mery D, Pedreschi F, Leon J (2006) Color measurement in L∗ a∗ b∗ units from RGB digital images. Food Res Int 39(10):1084–1091. https://doi.org/10.1016/j.foodres.2006.03.006

Luh BS, Kean CE (1988) Canning of vegetables. In: Kuh BS, Woodroof JG (eds) Commercial vegetable processing, 2nd edn. Van Nostrand Reinhold, New York

Lund DB (1982) Applications of optimization in heat processing. Food Technol 36(7):97–100

Mania I, Barone C, Caruso G, Delgado A, Micali M, Parisi S (2016a) Traceability in the Cheesemaking Field. The regulatory ambit and practical solutions. Food Qual Mag 03:18–20. ISSN 2336-4602

Mania I, Fiorino M, Barone C, Barone M, Parisi S (2016b) Traceability of packaging materials in the cheesemaking field. The EU regulatory ambit. Food packaging. Bulletin 25(4&5):11–16

Maijala R (2014) Scientific risk assessment-basis for food legislation. In: Ninios T, Lundén J, Korkeala H, Fredriksson-Ahomaa M (eds) Meat inspection and control in the slaughterhouse. Wiley, Chichester. https://doi.org/10.1002/9781118525821.ch26

McClane BA (2007) *Clostridium perfringens*. In: Doyle M. Beuchat L (eds) Food microbiology: fundamentals and frontiers, 3rd edn. ASM Press, Washington, DC, pp 423–444. https://doi.org/10.1128/9781555815912.ch19

Membré JM, van Zuijlen A (2011) A probabilistic approach to determine thermal process setting parameters: application for commercial sterility of products. Int J Food Microbiol 144(3):413–420. https://doi.org/10.1016/j.ijfoodmicro.2010.10.028

Mendoza F, Dejmek P, Aguilera JM (2006) Calibrated color measurements of agricultural foods using image analysis. Postharvest Biol Technol 41(3):285–295. https://doi.org/10.1016/j.postharvbio.2006.04.004

Mishkin M, Karel M, Saguy I (1982) Applications of optimization in food dehydration. Food Technol 36(7):101–109

Miyamoto K, Nagahama M (2016) Clostridium: food poisoning by *Clostridium perfringens*. In: Caballero B, Finglas PM, Toldrá F (eds) Encyclopedia of food and health. Academic Press, Oxford, pp 149–154. https://doi.org/10.1016/b978-0-12-384947-2.00171-9

NFPA/CMI Container Integrity Task Force (1984) Botulism risk from post-processing contamination of commercially canned foods in metal containers. J Food Prot 47:801–816

Nightingale RW, Stallings D (1986) Assessing the extra-commercial food needs of low-income countries. Food Policy 11(1):27–41. https://doi.org/10.1016/0306-9192(86)90045-x

Notermans SHW (1999) Clostridium—detection of neurotoxins of *Clostridium botulinum*. In: Encyclopedia of food microbiology, pp 463–466. https://doi.org/10.1006/rwfm.1999.0400

Olson KE, Sorrells KM (2015) 26. Thermophilic flat sour sporeformers. In: Salfinger Y, Tortorello ML (eds) Compendium of methods for the microbiological examination of foods. American Public Health Association, Washington, DC. https://doi.org/10.2105/mbef.0222.031

Palmieri A, Montanari A, Fasanaro G (2004) De-tinning corrosion of cans filled with tomato products. Corros Eng Sci Technol 39(3):198–208. https://doi.org/10.1179/147842204X2808

Parisi S (2004) Alterazioni in imballaggi metallici termicamente processati. Gulotta Press, Palermo

Parisi S (2012) Food packaging and food alterations: the user-oriented approach. Smithers Rapra Technology, Shawbury

Parisi S (2013) Food industry and packaging materials performance-oriented guidelines for users. Smithers Rapra Technology, Shawbury

Parisi S (2016) The world of foods and beverages today: globalization, crisis management and future perspectives. Learning.ly/ The Economist Group, London and New York. Available http://learning.ly/products/the-world-of-foods-and-beverages-today-globalization-crisis-management-and-future-perspectives. Accessed 29 June 2017

Parmee R (1990) X-rays give in-depth inspection. Sens Rev 10(2):84–86. https://doi.org/10.1108/eb007818

Pflung IJ, Odlaugh TE (1978) A review of z and F values used to ensure the safety of low-acid canned foods. Food Technol 2:63–70

Schmitt HP (1966) Commercial sterility in canned foods, its meaning and determination. Assoc Food Drug Off US Q Bull 30:141–151

Schoeninger V, Machado Coelho SR, Bassinello PZ (2017) Industrial processing of canned beans. Ciência Rural Santa Maria 4(05):e20160672. https://doi.org/10.1590/0103-8478cr20160672

Simpson BK (2012) Food biochemistry and food processing, 2nd edn. Wiley, New York

Teixeira AA (1971) Thermal process optimization through computer simulation of variable boundary control and container geometry. Dissertation, University of Massachusetts, Amherst

Tiwari BK, Singh N (2012) Pulse chemistry and technology. Royal Society of Chemistry, Cambridge

Upmann M, Bonaparte C (1999) Rapid methods for food hygiene inspection. In: Encyclopedia of food microbiology, pp 1887–1895. https://doi.org/10.1006/rwfm.1999.1320

White BL, Howard LR (2012) Canned whole dry beans and bean products. In: Siddiq M, Uebersax MA (eds) Dry beans and pulses production, processing and nutrition. Blackwell Publishing Ltd., Oxford. https://doi.org/10.1002/9781118448298.ch7

Yam KL, Papadakis SE (2004) A simple digital imaging method for measuring and analyzing color of food surfaces. J Food Eng 61(1):137–142. https://doi.org/10.1016/s0260-8774(03)00195-x